wmash 2004

Proceedings of the
Second ACM International
Workshop on Wireless Mobile Applications and
Services on WLAN Hotspots

Co-located with
MobiCom 2004

October 1, 2004

Philadelphia, Pennsylvania, USA

The Association for Computing Machinery
1515 Broadway
New York, New York 10036

Notice to Past Authors of ACM-Published Articles

ACM intends to create a complete electronic archive of all articles and/or other material previously published by ACM. If you have written a work that has been previously published by ACM in any journal or conference proceedings prior to 1978, or any SIG Newsletter at any time, and you do NOT want this work to appear in the ACM Digital Library, please inform permissions@acm.org, stating the title of the work, the author(s), and where and when published.

ISBN: 1-58113-877-6

Additional copies may be ordered prepaid from:

ACM Order Department
PO Box 11405
New York, NY 10286-1405

Phone: 1-800-342-6626
(US and Canada)
+1-212-626-0500
(all other countries)
Fax: +1-212-944-1318
E-mail: acmhelp@acm.org

ACM Order Number 103043
Printed in the USA

Message from the WMASH 2004 Organizing Committee Chairs

Welcome to the *Second ACM International Workshop on Wireless Mobile Applications and Services on WLAN Hotspots (WMASH'2004).* We take great pleasure in presenting you with a workshop on a very hot, timely, and challenging topic.

In a very rapid evolution of wireless networks, characterized by sudden rise and fall of technologies and solutions, the success of public WLAN Hot-Spots is continuing unabated. We recall that, last year, the first edition of this workshop was motivated by the attempt to tackle, in a timely manner, the emergence of the public WLAN phenomenon. This year, we are continuing to witness a great and massive interest from both the research community as well as from operators and hotspots deployers on public WLANs. We thus felt the need to maintain WMASH alive, and make it become a forum for the research community to convene and discuss technical and business challenges behind the evolution of WLAN from cable replacement to public access mean.

This year, WMASH has had a success well beyond our best expectations. The number of submissions has grown of about 50% with respect to last year's edition. We received a total of 47 submissions from 23 countries. The workshop program includes 15 excellent papers by authors from 7 countries, and many high quality submissions could not be accommodated for the space limitation. In addition to the presentation of accepted papers, the program also contains an exciting panel discussion on technical and business issues behind the evolution of WLAN hot-spots and hot-zones.

Behind a successful workshop, there is the tireless effort of many people which have smoothly and silently contributed to its organization. We would like to sincerely thank Dr. Parviz Kermani and Dr. Sung-Ju Lee, former organizers of WMASH 2003, for their expert suggestions and support "behind the scene." A depth of gratitude is owed to Dr. Anand Balachandran, our publicity chair, to Prof. Javier Gomez-Castellanos, our publication chair, and to Dr. Milind Buddhikot, our Panel Chair. They helped us with their invaluable talent and time to make this Workshop a reality. Many thanks is owed to the Technical Program Committee members and reviewers who thoroughly helped us select quality papers amongst many submitted manuscript. We had a well-balanced TPC with 28 members in total, with 16 from North America, 1 from Central America, 8 from Europe, and 3 from Asia. 14 members are in academia while 14 members are from the industry.

We hope you will find this workshop enjoyable, stimulating, worth to attend, and will contribute to its success by sharing your own ideas and discussing research and strategic directions for the public wireless Internet domain. It is our further hope that the success of this workshop at its second year motivates us and other to continue organizing it in the coming years.

The *WMASH 2004* Organizing Committee Chairs,

Giuseppe Bianchi	**Sunghyun Choi**	**Bill Schilit**
Univ. Roma Tor Vergata	*Seoul National University*	*Intel Research Lab*
Roma, Italy	*Seoul, Korea*	*Seattle, WA, USA*

Table of Contents

Author Index 147

WMASH'04 Organization

General Chair: Giuseppe Bianchi, *University of Roma Tor Vergata, Italy*

Technical Program Co-Chairs: Sunghyun Choi, *Seoul National University, Korea*
Bill Schilit, *Intel Research, Seattle, USA*

Publicity Chair: Anand Balachandran, *Microsoft, USA*

Publications Chair: Javier Gomez, *National University of Mexico, Mexico*

Panel Chair: Milind Buddhikot, *Lucent Technologies, USA*

Technical Program Committee: Victor Bahl, *Microsoft Research*
Anand Balachandran, *Microsoft*
Milind Buddhikot, *Bell Labs, Lucent Technologies*
Andrew Campbell, *Columbia University, New York*
Marco Conti, *CNR - Instituto CNUCE*
Gabor Fodor, *Ericsson*
Rosario Garroppo, *University of Pisa*
Javier Gomez, *National University of Mexico*
Xingang Guo, *Intel Corporation*
Jeffrey Hightower, *University of Washington, Seattle*
Kyungghun Jang, *Samsung Advanced Institute of Technology*
Byoung-Jo Kim, *AT&T Labs - Research*
Young-bae Ko, *Ajou University*
John Krumm, *Microsoft Research*
Sung-Ju Lee, *HP Labs*
Saishankar Nandagopalan, *Philips Research, USA*
Maria Papadopouli, *U. of North Carolina, Chapel Hill*
George Polyzos, *Athens University of Economics & Business*
Anand Prasad, *DoCoMo Euro-labs*
Daji Qiao, *Iowa State University*
Ramachandran Ramjee, *Bell Labs, Lucent Technologies*
Puneet Sharma, *HP Labs*
Rajeev Shorey, *IBM Research, India*
Raghupathy Sivakumar, *Georgia Institute of Technology*
Ilenia Tinnirello, *University of Palermo*
Iakovos Venieris, *National Technical University of Athens*
Yang Xiao, *University of Memphis*
Michael Zorzi, *University of Ferrara*

External Reviewers: Vangelis Angelakis   Francesco Oppedisano

Leonardo Badia   Konstantina Papagiannaki

Nicola Bonelli   Michele Rossi

Raffaele Bruno   Manolis Spanakis

Felix Hernandez-Campos   Luca Tavanti

Stefano Lucetti   Nikos Tselikas

Gaia Maselli   Giovanni Turi

Pietro Michiardi   Jing Zhu

Vaggelis Nikas

Sponsored by:

SOWER: Self-Organizing Wireless Network for Messaging

Mark Felegyhazi, Srdjan Čapkun, Jean-Pierre Hubaux

Laboratory of Computer Communications and Applications,

EPFL – Switzerland,

email: {mark.felegyhazi,srdan.capkun,jean-pierre.hubaux}@epfl.ch

ABSTRACT

Short Message Service (SMS) has become extremely popular in many countries, and represents a multi-billion dollars market. Yet many consumers consider that the price charged by the cellular network operators is too high. In this paper, we explain that there exist alternatives to cellular networks for the provision of SMS. In particular, we present the Self-Organizing Wireless messaging nEtwoRk (SOWER), an all-wireless network operable in cities. In SOWER, each user installs a wireless, power-plugged device at home and communicates by means of a mobile device. Based on our experimental measurements of IEEE 802.11 equipped devices, we show the feasibility of the concept in various urban scenarios. We also show that city-wide connectivity can be achieved even with a limited market penetration. We explain that the capacity of such networks is sufficient to support messaging communication.

Categories and Subject Descriptors

C.2.1 [**Computer-Communication Networks**]: Network Architecture and Design—*Wireless communication*; C.2.2 [**Computer-Communication Networks**]: Network Protocols—*Applications*

General Terms

Design, Performance

Keywords

Ad hoc Networks, Messaging, Self-organization, Distributed Computing

1. INTRODUCTION

The operation of cellular networks is by far the largest business segment of mobile networking. In these networks, voice service is still the dominant source of revenue. Data

services, however, are becoming more and more widespread; in particular, short message services (SMS) have become extremely popular. According to [31], they will account for a total revenue of $84 billion of the cellular operators in 2008.

In spite of the fact that the provision of this service is relatively straightforward for the operators, in many countries the price of SMS usage is fairly high; consumer associations reproach the operators that they are taking advantage of their oligopolistic situation to maintain outrageous price rates [32]. The operators respond that the cost of their overall infrastructure must be taken into account in the computation of a "fair" pricing scheme.

Much of the controversy is due to the absence of an alternative for the end users. In this paper, we will explain that this situation is about to change: alternatives to cellular networks are becoming available for the provision of messaging services; an important characteristic of these alternatives is that they can be partially or totally operated by the end users.

A first alternative consists in having a large number of end users open the access to their home-based Internet-connected WLAN access points (APs); in this way, mobile users passing by can connect to the APs to send and receive messages. The nice property of this solution is that it removes charging per message, as the connection of the AP to the Internet is usually charged at a flat rate; the amount of the messaging traffic would be negligible considering the available bitrate. But there are drawbacks: the operators of the Internet access may forbid (for example, for security reasons) this kind of open access; in addition, the person of the household managing the access point and paying for the Internet access subscription may be unwilling to open his personal communication and computing infrastructure to the SMS users of the neighborhood. Hence, the service availability provided by this alternative might be limited.

In this paper, we explore an all-wireless alternative: we show that a city-wide short message service can be supported by a user-operated wireless network, without using even a single wireline access. To our best knowledge, this solution was never studied so far.

The solution we propose is based on a two-tier architecture: a user is expected to install a fixed, power-plugged device at home, and to communicate via a hand-held device; the home device keeps track of the location of the mobile one. The home devices organize themselves to set up a *wireless backbone* over which they transmit the messages. Each mobile device is attached to the wireless backbone via a nearby home device. Of course, as we will see, if some home devices have also wireline connectivity to the Internet, this can only increase the performance of SOWER.

To demonstrate the feasibility of the solution, we report real connectivity measurements between laptops equipped with IEEE 802.11 cards, that we have performed in a city. Based on the measurements, we estimate the device density and the market penetration required to reach network connectivity in various urban scenarios. Furthermore, we explain why capacity, well-known to have a major scalability problem in ad hoc networks, is sufficient to support messaging communication. Based on current market prices, we show in Section 7.2 that the proposed solution is cost-effective for the frequent SMS users.

The remainder of the paper is organized as follows. In Section 2, we give an overview of related work. In Section 3, we provide the description of our proposal for messaging networks. In Section 4, we present our results of connectivity and coverage in realistic scenarios. In Section 5, we investigate capacity issues. More technical issues are presented in Section 6. We discuss the deployment of the network along with additional business issues in Section 7. Finally, we conclude our paper in Section 8.

2. RELATED WORK

In this section we overview the existing wireless network architectures and we discuss new wireless networks proposed in the literature.

2.1 Existing network architectures

Cellular networks, such as GSM networks, are the most prominent examples of existing wireless architectures [23]. These networks are typical examples of networks with *pre-deployed* infrastructure. Using a complex infrastructure, the network coverage can be almost 100 per cent in a given area. Cellular networks were originally designed to carry voice traffic, but the recent evolution towards new generations of cellular networks also enables data communication besides traditional voice communication. The communication rate for both voice and data communication is low, compared to the data rate in computer networks.

Another example of existing wireless networks are paging networks [9]. These networks are used to provide one-way communication with short messages. The messages are used to notify the users, but users cannot reply to the notification. Like cellular networks, paging networks also rely on a pre-deployed infrastructure. Paging networks provide an almost full coverage as well. It is possible to have two-way communication with a paging system, but this requires a more sophisticated mobile device or a PC. Due to the messaging communication, the traffic rate is extremely low compared to that of computer networks.

A different example of wireless networks is a Wi-Fi network [24]. These networks differ from cellular and paging networks in that they typically provide a coverage limited to a few access points, but with broadband access to the Internet. These *hot spots* are deployed in designated places, such as airports and train stations. Because of the limited coverage of the Wi-Fi network, users cannot change their place during the communication session. Current development efforts, such as the white paper [33], address the problem of seamless mobility in Wi-Fi access networks.

2.2 Proposed networks in the literature

Recently, novel wireless architectures have been proposed in the literature. Their main difference with respect to existing wireless networks is that they no longer depend on a pre-deployed, centrally managed infrastructure, but they operate in a self-organized manner. The most prominent example of these networks are ad hoc networks. In ad hoc networks, mobile devices perform all networking tasks (i.e., routing, packet forwarding) in a self-organized manner without relying on an existing infrastructure. The advantage of ad hoc networks is that they can be easily deployed at a low cost; the disadvantage is that they do not scale for all types of traffic. Notably, it has been shown in the work of Gupta and Kumar [8] that the capacity per user diminishes as the network size increases and that: "... scenarios envisaged in collections of smart homes, or networks with mostly close-range transactions and sparse long-range demands, are feasible." In spite of this last sentence, the capacity problem described by the authors has been often interpreted as a result that shows the infeasibility of large scale ad hoc networks in general.

To overcome this problem, researchers proposed to combine ad hoc networks with existing infrastructures, such as cellular networks. This integration of cellular and ad hoc networks results in *hybrid ad hoc networks*. In [17] Lin and Hsu present a multi-hop cellular architecture to extend the coverage of existing cellular system. Luo *et al.* describe a Unified Cellular and Ad Hoc Network (UCAN) framework in [19] that enhances control area throughput while maintaining fairness. The authors propose a fair 3G cellular base station scheduling protocol, an access discovery mechanism and a secure crediting system. In [28], Wu *et al.* present an Integrated Ad Hoc Cellular Relaying System (iCAR). The iCAR system can efficiently balance traffic loads between cells of the cellular system by using ad hoc relaying stations.

Several researchers proposed to replace the flat ad hoc network architecture with a hierarchical architecture. In the case of a hierarchical ad hoc network, there exists a backbone of wireless devices that have more powerful computation and transmission capabilities. Thus, they can relay the traffic coming from low-tier devices. Networking protocols can exploit the hierarchical structure of the network: Instead of a flat routing architecture, one can propose a clustered routing scheme (e.g., the solution proposed in [29]). In [16], Karrer, Sabharwal and Knightly propose a network based on pre-deployed Transit Access Points (TAPs), that serve as a high-speed, multi-hop, wireless backbone with a limited number of access points to the Internet. In [14], Jetcheva *et al.* propose Ad Hoc City, a city-wide, multi-tier ad hoc network based on vehicles. Small and Haas [26] describe an Infostation model (SWIM) that is based on the capacity-delay tradeoff of ad hoc networks. They demonstrate their solution on a system used to observe the behavior of whales.

2.3 Signal propagation

In Section 4, we will investigate connectivity in SOWER. Since connectivity depends on the radio propagation in the given environment, we briefly review the state of the art in this field.

Radio propagation has been extensively studied for cellular networks (for a comprehensive overview, see [23]). There exist several propagation models in urban environments for outdoor and also for indoor scenarios.

For outdoor signal propagation in urban areas, a widely used model is the Okumura-Hata model [12]. Unfortunately, neither this model nor its extension up to 2 GHz give precise propagation results for personal communication systems that have a communication range less than 1 km.

For indoor radio propagation, researchers use attenuation models derived from experiments. It is more difficult to define an appropriate propagation model for indoor than for outdoor, because propagation depends very much on the particular characteristics of the indoor environment. A detailed study of indoor propagation models is presented by Hasemi [10] and more recently by Hassan-Ali and Pahlavan [11]. Signal propagation and the effect of interference were also studied for the IEEE 802.11 system (e.g., [15]).

In the literature, there is currently no unified model that describes both outdoor and indoor radio propagation. In particular, there exist no analytical model to describe signal propagation from outdoor to indoor environment and vice versa. Propagation loss into buildings is determined by several factors, such as the number of windows, the material of the building and the vertical distance of the receiver from the sender.

3. AD HOC MESSAGING NETWORK

In this section we present our solution and discuss its properties in more detail.

3.1 System description

In this section, we propose a novel application scenario for ad hoc networks, which we call a *Self-Organizing Wireless messaging nEtwoRk (SOWER)*. We present an example of SOWER with a message transmission in Figure 1.

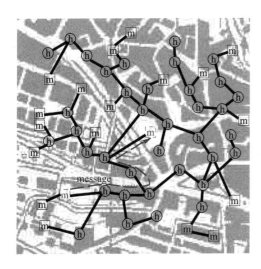

Figure 1: An example of SOWER: The network consist of a set of home devices (h) and mobile devices (m). The dashed line represents the route of a multihop message forwarding.

SOWER consists of two types of devices: a set of mobile devices and a set of static devices. We assume that each user owns two devices, one of each type. The user makes use of the first device as a hand-held device that we call a *mobile device*. We assume that the mobile device is powered by a rechargeable battery. We further assume that the user sets the second device at a fixed place connected to a permanent power source. Thus, we call the static device a *home device*. The deployment place for the home device is typically the home or the office of the user.

We make the following assumptions: All devices have similar radio capabilities. If two devices reside within the transmission range of each other, then they are considered to be neighbors. The devices periodically perform a neighbor discovery procedure, and are aware of their neighborhood. The radio links between neighbors are assumed to be bidirectional. All devices operate in the same license-free ISM frequency band (e.g., at 2.4 GHz). Both mobile and home devices are equipped with a radio card that enables ad hoc networking. The devices rely on the CSMA/CA medium access method (e.g., IEEE 802.11b technology). Note that our proposal can rely on other technologies as well.

The devices form a wireless ad hoc network. The purpose of the ad hoc network is to provide *messaging communication* between the users. We call *message* the unit of user information that can be transmitted in a single packet (e.g., an SMS message in cellular networks or an email on the Internet). Message transfer between distant devices may involve multiple wireless hops. In our architecture, the mobile devices rely on the home devices to relay the traffic. We refer to the connected set of home devices as a *wireless backbone*. We assume that a routing protocol is implemented in each home device to transfer packets from the source to the destination. We also assume that a substantial amount of the traffic is limited to the city covered by the network.

In our proposal, we assume that the network is under the full control of the users, meaning that no central authority supervises the operation of the network. We will relax this assumption in Section 7. We assume, of course, that there exist companies that produce the wireless devices.

4. CONNECTIVITY AND COVERAGE

In this section, we analyze the connectivity and coverage of SOWER in city scenarios. Our goal is to assess the required density of home devices in order to have a connected network[1]. We first present field test results for connectivity in a city scenario for radio propagation. Based on these results, we perform an extensive simulation study of connectivity in two-dimensional city scenarios. Finally, we extend our connectivity investigations to a three-dimensional area.

4.1 City scenario - Field test measurements

Although radio range is often modelled with a circle, this is obviously not an appropriate model in urban environments. None of the existing models presented in Section 2.3 can be used for our problem. Motivated by the lack of analytical results, we decided to perform a large number of connectivity measurements in a small city, namely in Lausanne, Switzerland. For this purpose, we used laptops equipped with Cisco Aironet [34] wireless cards. All wireless cards were compliant with the IEEE 802.11b standard. We operated the cards with a power of 100 mW, with the transmission rate set to "auto-rate selection" between 1 and 11

---

[1]The connectivity could be further increased by taking the mobile devices also into consideration. But we refrain from doing it, because it would result in higher power consumption on these battery-operated devices.

Mbit/s. We measured different types of links where we put the laptops indoors and outdoors.

We randomly chose measurement points in the center of the city to represent different types of links (as shown in Figure 2). We performed most of the tests in the streets and at the ground floor of the buildings. We also made some "vertical" measurements, meaning across the floors of the buildings. In our measurements, most of the devices reside indoor, which is compliant with the operating principles of SOWER.

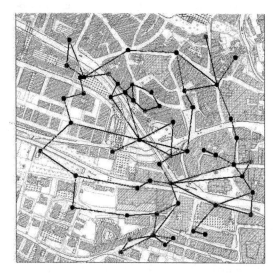

Figure 2: Map of the downtown area (500m * 500m) of Lausanne with our connectivity measurement results. In our measurements, most of the devices reside indoor that is compliant with the properties of SOWER. (© Service du cadastre de la ville de Lausanne)

We identified different types of links that depend on the position of the endpoints and the type of propagation medium. The connectivity results for different types of possible links are shown in Table 1. The first two columns show the type of the link. We present the number of measurements in the third column. Columns 4 and 5 summarize the average value and standard deviation of the test results for each link type. The last column presents the simulated interval for radio range that is derived from the measurement results. The rows represents different link types. Note that the connectivity is affected by the material of the obstacles as well[2].

Our measurement results also show that the small city center such as the one in Lausanne can be covered with a small number of devices. In our city, a network of approximately 55 devices are enough to cover the 500m * 500m city center.

From our measurements in Lausanne, we generalize our results to a metropolis and a suburban area using simulations. Although the structure of buildings is different in a metropolis, a small city or a suburban area, there are common characteristics, such as the type of windows and doors. Radio signals propagate mainly through these light-weight

---

[2]Special material can break down the connectivity. In one of our measurements, the signal was not able to go through two special window glasses, but it could easily go through one.

elements in a building, making the generalization justified. In our future work, we intend to pursue a more extensive measurement campaign in different scenarios.

4.2 City scenario - Simulation results

We performed simulations using three different urban scenarios to assess connectivity in a two-dimensional area. We set the parameters of the generic model to represent three realistic city scenarios: (i) a metropolis with large buildings and wide avenues, (ii) a small city center with small streets and (iii) a suburban area including houses and open space. We defined the radio range in the simulations from the field-test measurements described in Section 4.1. All simulation results are the average of 100 runs with a confidence interval of 95%.

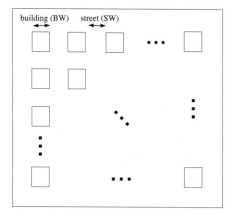

Figure 3: City scenario for connectivity and coverage experiments.

We used a simplified general city setting as shown in Figure 3. In the simulations, we generate a symmetric scenario of a total area of 1 km^2 with a given street width (SW) and building width (BW). Table 2 summarizes the parameter values for the city scenarios. Our parameter settings correspond to the standard street width in urban street planning (e.g., [27]). The standard street size follows the standard design technique for city planning called *Transit Oriented Development (TOD)* [4]. In all simulations, we uniformly distribute the home devices among the buildings. We denote the *density of the home devices* by δ throughout the paper and we express δ in devices/km^2.

Parameter	Metropolis	Small city	Suburban area
SW	40 m	20 m	35 m
BW	80 m	50 m	13.25 m

Table 2: Parameter values for the city scenario coverage simulations.

In all cases, we consider the largest connected component of the set of home devices. We investigate the following performance measures:

1. **Coverage of the largest connected component (denoted by κ):**

$$\kappa = \frac{A_l}{A}$$

4

Scenario (name)	Scenario (pictogram)	Number of measurements	Average range	Standard deviation	Simulated range (Section 4.2)
O-O-O		12	124.22 m	32.22 m	–
O-I-O		13	70.3 m	20.6 m	–
I-X-O		26	77.24 m	32.64 m	50-100 m
I-O-I		14	71.75 m	27 m	50-100 m
I-I-I		12	38 m	15.82 m	25-50 m

Table 1: **Connectivity measurements in a downtown scenario for different link types. "I" stands for indoor, "O" stands for outdoor and "X" stands for any of the two environments. The link A-B-C means: A - the type of the first end of the link, B - the type between the endpoints, C - the type of the second end of the link.**

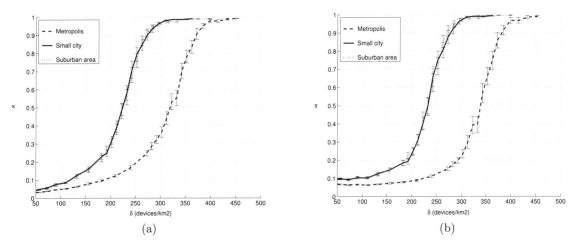

(a) (b)

Figure 4: **Coverage and connectivity in the two-dimensional network; (a) coverage of the largest connected component of home devices (κ); (b) fraction of the size of the largest connected component and the total number of home devices (π).**

where A_l is the area covered by the largest connected component of home devices and A is the total simulation area.

2. **Fraction of the number of devices in the largest connected component over the total number of devices (denoted by π):**

$$\pi = \frac{n_l}{n}$$

where n_l is the number of home devices in the largest connected component and n is the total number of home devices.

The connectivity for a given connection type is calculated from the measurement results presented in Section 4.1: The radio range for a device is a uniform random variable between the extreme values of that link type. This characterizes the fact that propagation loss varies with the material of the buildings.

First, we present our results for the coverage of the home devices. Figure 4a shows the value of κ as a function of the device density (δ) for all three scenarios.

We can observe that the coverage requirement for a small city and a suburban area is almost the same. In both scenarios, we can reach a coverage near to 100% with approximately 300 devices. In the metropolis, approximately 400 devices are needed to reach the full coverage. The full coverage is only possible if the network becomes almost fully connected. Figure 4b presents the value of π as a function of δ.

The two-dimensional results are relevant in the suburban scenario and in the small city scenario with low buildings. They are less relevant in the metropolis scenario, because in that case the network expands in three dimensions. To present more significant results for the metropolis scenario, we investigate the properties of a three-dimensional network in the following subsection.

5

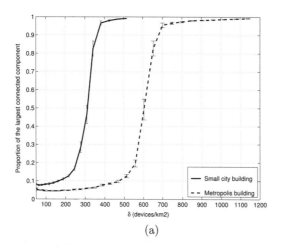
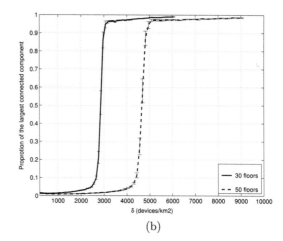

| (a) | (b) |

Figure 5: Fraction of the size of the largest connected component over the total number of home devices in a three-dimensional network; (a) in small buildings; (b) in skyscrapers.

Scenario	small	historic	modern	ultra-modern
Population density in persons/km²	2260	8177	12500	25850
Device density requirement in devices/km²	380	700	3000	5000
Market penetration requirement from simulations (power = 100mW)	0.168	0.086	0.24	0.193
Calculated market penetration requirement (power = 1W, α = 5)	0.04	0.02	0.06	0.05

Table 3: Device density requirements vs. population density in different scenarios.

4.3 Three dimensional network

In this section, we investigate the connectivity results for three-dimensional networks in the small city and metropolis scenarios.

According to our field tests, coverage in a building is typically 25-50 meters (depending on the environment), if both endpoints of the link reside on the same floor. Vertical coverage differs from the coverage on the same floor, because of the structure of the wall that is between the floors in a building. In our field test, we measured signal propagation in the vertical direction as well. Our test results show that a three-dimensional link I-I-I can cover 1 to 3 floors in a building if there are open areas between the floors, like moving stairs etc. According to our measurement results, considering an office building with very little open space between the floors, the vertical coverage of the link is at most one floor (e.g., 3-7 meters). If the devices are in a shopping center with big open areas and a big moving stairs, then the vertical coverage extends up to 3 floors (12-15 meters).

We present our simulation results for connectivity in the three-dimensional case[3]. We consider a small city scenario with buildings of 5 floors and three subtypes of the metropolis scenario, where buildings consist of 5, 30 and 50 floors, respectively. We randomly choose a building for each device with a uniform probability and we also determine a uniformly random position within the building. We consider a three-dimensional network, where we extend the notion of I-O-I links for adjacent floors in adjacent buildings[4].

Figures 5a and 5b show connectivity results for small buildings and skyscrapers, respectively. In all cases, the simulation area is 1 km². The connectivity, and therefore the coverage, of the network increases significantly if δ is above a certain threshold.

Table 3 summarizes the device density requirements in the considered scenarios[5]. As a comparison, we present population density data from cities that represent the small city and three subtypes of the metropolis scenario, respectively (for the original data, see [35]):

- Small city center (e.g., Berkeley, California, USA) - small buildings with 5 floors

- Historic metropolis center (e.g., Rome, Italy) - large buildings with 5 floors

- Modern city center (e.g., Berlin, Germany) - large buildings with 30 floors

- Ultra-modern city center (e.g., Manhattan, New York City, USA) - large buildings with 50 floors.

Note that the values presented in Table 3 express a pessimistic approximation of the requirements for device density. Several conditions can accelerate the deployment of SOWER, as we discuss in Section 7.

We note here that in all our measurements, we operated the IEEE 802.11 adapters at 100 mW, as this is the maximum power allowed in the considered city. In several countries (including the USA), the maximum transmission power

---

[3]For a large number of devices (i.e., if the number of devices increases over 2000), the simulation of coverage becomes infeasible. Hence, we present only connectivity results.

[4]This means that a device located for example at floor 5 can communicate with devices that reside in floors 3, 4, 5 in an

adjacent building, if they are within the given communication range defined for I-O-I links

[5]Market penetration expresses the fraction of users in the network and the total population.

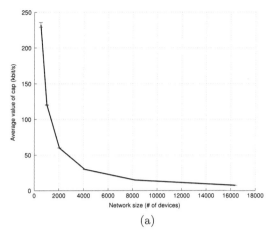 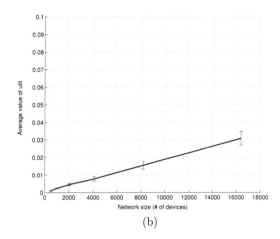

(a)   (b)

Figure 6: Capacity simulations: (a) Average value of capacity per home device (cap); (b) Average channel utilization per home device ($util$).

for IEEE 802.11 wireless card is 1 W. As a results, the network can operate with a lower density of users as shown in Table 3. Interference affects connectivity as well; we discuss this issue in Section 5.

5. CAPACITY

In this section, we show that the capacity of SOWER is high enough to carry messages in an all-wireless network. We define *capacity* for a home device as the maximum available throughput.

We investigate the following performance measures:

1. **Capacity per home device:**

$$cap = \frac{link\ capacity}{number\ of\ neighbors}$$

   where link capacity means the maximum possible throughput of the radio card in an isolated environment (i.e., if there are no other transmissions and there is no interference).

2. **Channel utilization per home device:**

$$util = \frac{traffic\ load}{cap}$$

   where traffic load means the aggregate traffic that the home device has to transmit.

This notion of capacity per home device is meaningful, if the channel is not saturated, notably because there is a small probability of collisions. If the channel saturates, then the definition of throughput becomes more complicated. In the saturation case, the results in [1] can be applied.

In order to investigate capacity on the wireless backbone, we developed a city simulator in C. We first simulate a test network of 500 devices on an area of 1 km^2. We distribute the devices according to the assumptions presented in Section 4. We assume that the users move according to the Gauss-Markov mobility model [3] with a randomness factor of 0.9; this approach provides a better model for pedestrian

movements in a city than the commonly used random waypoint mobility model. We assume that all devices have a radio range of 75 meters, which is approximately the average radio range obtained from our field test results. We assume that a routing protocol is running on the wireless backbone as presented in Section 6.1. We assume a heavy traffic load (e.g., during rush hours) with a message sending rate of 1 message per minute. We assume that the user message length is 256 bytes (which corresponds to the size of an SMS message in GSM networks) and the length of a location update packet is 40 bytes. We increase the number of devices (and therefore the device density δ) exponentially from 512 to 16384 devices. As shown in Figure 6a, the maximum available channel capacity is inversely proportional with δ. Because the set of home and mobile devices increases equally as δ increases, the traffic load per home device remains constant. As a result, the channel utilization per home device increases linearly with δ, as presented in Figure 6b.

Next, we perform a simulation to investigate the effect of the network size on the traffic load. We increase the size of the network exponentially from 16 to 16384 devices; we also increase the size of the simulation area from 125m x 125m to 4km x 4km to keep the density δ equal to 1024 devices/km^2. Figure 7 shows the average traffic load at each home device with increasing network size. Our results show that the average traffic load increases proportionally with $\sqrt{n}$ as the network size increases.

The simulation results show that channel utilization is low even for large networks. The average traffic load per home device is some orders of magnitude lower than the capacity per home device. As a result, the links are generally not congested. Even if a link becomes temporarily congested, the "store-and-forward" principle of message transmission enables to store messages until this transient situation ends.

It is important to mention that we do not take the effect of collisions and interferences into account. As we have shown in this section, the traffic rate is very small compared to the available maximum throughput. Thus, collisions occur rarely and their effect is negligible. As the data rate is low, the effect of interferences within the network is not very relevant either. The effect of possible interfer-

7

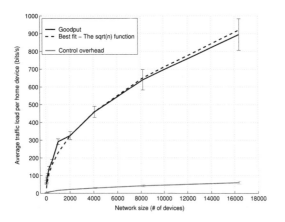

Figure 7: Average traffic load per home device as a function of network size.

ences due to other networking technologies operating in the same ISM frequency band is more significant. But, due to the delay-tolerant property of the communication, messages can be stored at intermediate devices until the connection is restored. For a detailed analytical study on the effect of interference, the reader is referred to [5] and [25]. Due to these effects, the real capacity of SOWER is smaller than our simulation results. Let us emphasize, however, that we assumed a common link capacity of 1 Mbit/s. Current IEEE 802.11 technologies enable communication with a bitrate up to 54 Mbit/s. Thus, the available throughput can be much higher than the one we consider in our investigations.

6. ADDITIONAL TECHNICAL ISSUES

In this section, we address three issues that are also important in SOWER.

6.1 Routing

We assume that the device density is high enough to enable the home devices to be connected. We assume that a distance vector routing algorithm is running on the wireless backbone network (for example DSDV [22]). Due to the static network configuration, the overhead of global broadcasts is small. We further assume that the message routing between mobile nodes is based on the MobileIP scheme [21] where home devices correspond to the home agents for their mobile and they are also the foreign agents for other mobiles. Due to space constraints, we provide the detailed description of the routing protocol in [7].

6.2 Access to the infrastructure and charging

If the device density is high enough in SOWER, then it can operate in a city without accessing an existing infrastructure. In this case, message forwarding can be performed within the network without charging. However, if the device density is not high enough or the network expands to several cities, SOWER can coexist with existing wireless networks, such as a cellular system. We assume, according to the traditional approach, that the sender of the message is charged if the message is transmitted to the destination using a cellular or a Wi-Fi network.

Due to the MobileIP-based routing scheme, users are naturally motivated to keep their home devices turned on. But a user my be tempted to modify the behavior of her home device to inhibit the relay function, as she does not obtain a benefit from it. In order to motivate users to refrain from doing so, a "virtual money" can be introduced (e.g., as described in [2, 30]).

6.3 Security

Security is also a crucial issue in SOWER. It may be necessary to provide confidentiality and integrity of the messages. To fulfill these requirements, each message should be encrypted between the sender and the receiver using well-established cryptographic techniques. In particular, the location update procedure of the routing scheme has to be secured to prevent attackers to send false location updates and redirect all messages of a given users to themselves.

A more challenging problem is how to secure routing, in order for example to prevent a malicious user from dropping or redirecting messages. The problem of secure routing has been extensively studied in recent years (e.g., in [13, 20]). The choice and adaptation of the most appropriate proposal is left for future study.

Privacy is also an important question in SOWER. The mobile users should be protected from malicious location tracking.

The issue of trust is closely related to security. Without the presence of an operator, the users have to maintain the network themselves. The operation has to be based on the emergence of mutual trust. Trust can be incorporated into the system as a reputation mechanism for example [18].

6.4 Addressing

In SOWER, a distributed addressing solution is needed to route messages correctly. Recently, several papers proposed addressing schemes for ad hoc networks. In particular, Eriksson, Faloutsos and Krishnamurthy describe a scalable dynamic addressing scheme in [6] that can be applied to SOWER as well. The solution is based on the separation of the node's identifier from the routing address.

7. DISCUSSION

This section discusses practical aspects related to the initial deployment and to the business issues.

7.1 Deployment of the network

If one aims at the realization of a small scale network, like close communication among friends in a given area, then the deployment of the network is fast. Especially in small towns or villages, social groups tend to be close in terms of distance as well. If the envisioned network is large, then additional solutions are required to ensure communication until the device density reaches a satisfactory level.

A possibility for overcoming the problem of insufficient connectivity in the network is to rely on the existing infrastructure to carry the messages.

- If no connectivity is available with the destination, the devices can exploit an access to a cellular network. In this case, the users should be notified that they will be charged by the operator of the infrastructure network.

- High speed Internet connections are more and more popular. According to [36], in October 2003 in the U.S., 60% of the households had an Internet connection and according to [37] 40 % of them were high-speed. Internet connections provide another solution to connect unconnected areas.

The deployment of the network can be faster if a manufacturer provides dual-mode devices. Such a device could be a cellular mobile phone that is able to operate in ad hoc mode as well. There are several products that already provide access to both cellular networks and WLANs. We mention a subset of them such as: the Sierra Wireless AirCard 555 [38], the Nokia GPRS / WLAN card - (D211 / D311) [39] and the GlobeTrotter COMBO PCMCIA card for WLAN / GPRS / GSM [40].

7.2 Business issues

In this subsection, we derive a financial motivation of the users. We base our study on data acquired for the United Kingdom. We use the real data to justify the need for free message sending from the user's point of view.

The population of UK is 59 million. As mentioned in [41], the mobile penetration in the UK is about 0.74 (the number of registered users is 43.5 million). According to [42], there has been 20 billions of SMS sent in 2003.

Parameter	Frequent senders	Rare senders
Number of users	8.7 millions	34.8 millions
SMS sent/year	16 billions	4 billions
SMS/day/user	5	0.315
SMS/year/user	1839	115
price/year/user	$334 / £184	$22 / £12

Table 4: SMS data in the UK for 2003

We assume that Pareto's 80-20 rule applies to SMS messaging, namely that 20 per cent of the users send 80 per cent of the messages. In general, the price of an SMS message is 0.10 pounds. Thus, we can calculate the SMS requirements in terms of both number of SMS messages sent and price paid for each type of users (i.e., frequent senders and rare senders) as shown in Table 4. We see that for frequent users, the SMS costs in one year can easily cover the costs of buying a mobile and a home device. Thus, our proposal is cost-effective for them.

7.3 Operators' strategy

Clearly, the prospect of SOWER can be perceived as a threat by cellular operators, as it is susceptible to jeopardize the revenue obtained from the frequent SMS users. However, it can also be an opportunity: the operators can try to surf on this potential new fad by being the enabling company that deploys the first "home" devices to bootstrap connectivity; they would in any case remain the unavoidable solution for long-range (meaning inter-city) connectivity. This strategy would be similar to the one adopted by several incumbent operators with respect to the deployment of hot spots.

8. CONCLUSION

In this paper, we have studied SOWER, an all-wireless, user-operated alternative to cellular networks for the provision of SMS in urban environments; we relied as much as possible on the real data of propagation measurements and of city topology. The conclusions are very encouraging: (i) the city-wide connectivity can be achieved even with a modest market penetration, in all the city scenarios we have studied; and (ii) the capacity is sufficient to support messaging, owing to the modest bandwidth needs of SMS with respect to the high bitrate of the wireless links.

We believe that the feasibility (and possibly the real deployment of SOWER in some cities) will have a beneficial influence on the pricing of SMS: this prospect will have an impact on the pricing of messaging similar to the impact that Voice over IP has had on the pricing of conventional voice service.

In terms of future work, we intend to pursue our measurement campaign in order to corroborate our simulation results in specific scenarios. We will further study the charging and security issues discussed in Section 6. Finally, we intend to implement a routing protocol and study its behavior in a prototype setting.

9. ACKNOWLEDGEMENTS

The work presented in this paper was supported (in part) by the National Competence Center in Research on Mobile Information and Communication Systems (NCCR-MICS), a center supported by the Swiss National Science Foundation under grant number 5005-67322 (http://www.mics.org).

We would like to thank to Prof. Edward Knightly for his comments. We are also thankful to Gilles Cherix and Cedric Gaudard, who performed the connectivity measurements in Lausanne.

10. REFERENCES

[1] G. Bianchi, "Performance analysis of the IEEE 802.11 distributed coordination function," *IEEE Journal on Selected Areas in Communications*, 18(3):535-547, March 2000.

[2] L. Buttyan and J. P. Hubaux, "Stimulating Cooperation in Self-Organizing Mobile Ad Hoc Networks," *ACM/Kluwer Mobile Networks and Applications (MONET)*, October 2003, Vol. 8 No. 5

[3] T. Camp, J. Boleng, and V. Davies, "A Survey of Mobility Models for Ad Hoc Network Research," *Wireless Comm. and Mobile Computing (WCMC): Special Issue on Mobile Ad Hoc Networking: Research, Trends, and Applications*, vol. 2, no. 5, pp. 483-502, 2002.

[4] H. Dittmar, G. Ohland (Editors), *The New Transit Town: Best Practices in Transit-Oriented Development*, Island Press, 2003.

[5] O. Dousse, F. Bacelli, P. Thiran, "Impact of Interferences on Connectivity in Ad Hoc Networks," *Proceedings of IEEE Infocom 2003*, San Francisco, USA, March 30 - April 3, 2003.

[6] J. Eriksson, M. Faloutsos, and S. V. Krishnamurthy, "Scalable Ad Hoc Routing: The Case for Dynamic Addressing," in *Proceedings of IEEE INFOCOM 2004*, Hong Kong, March 7-11, 2004.

[7] M. Felegyhazi, S. Čapkun, J.-P. Hubaux, "SOWER: Self-Organizing Wireless Network for Messaging," EPFL, technical report, Nr: IC/2004/62, July, 2004

[8] P. Gupta and P. R. Kumar, "The Capacity of Wireless Networks," *IEEE Transactions on Information Theory*, pp. 388-404, vol. IT-46, no. 2, March 2000.

[9] L. Harte, *An Introduction to Paging Systems*, Althos Publishing, 2004.

[10] H. Hasemi, "The Indoor radio Propagation Channel," *Proceedings of IEEE* vol. 8, No. 7, pp. 943-968, July 1993.

[11] M. Hassan-Ali, K. Pahlavan, "A New Statistical Model for Site-Specific Indoor Radio Propagation Prediction Based on Geometric Optics and Geometric Probability," *IEEE Transactions on Wireless Communication*, 1(1), pp. 112-124, Jan 2002

[12] M. Hata, "Empirical Formula for Propagation Loss in Land Mobile Radio Services," *IEEE Transactions on Vehicular Technology*, vol. vt-29 No. 3, pp. 317-325, August 1980.

[13] Y.-C. Hu, A. Perrig, and D. B. Johnson, "Ariadne: A Secure On-Demand Routing Protocol for Ad Hoc Networks," *Proceedings ACM MobiCom 2002*, pp. 12-23, ACM, Atlanta, GA, September 2002.

[14] J. G. Jetcheva, Y.-C. Hu, S. PalChaudhuri, A. K. Saha, D. B. Johnson, "Design and Evaluation of a Metropolitan Area Multitier Wireless Ad Hoc Network Architecture," *Proceedings of IEEE WMCSA 2003*, IEEE, Monterey, CA, October 2003

[15] A. Kara, H. L. Bertoni, "Blockage/Shadowing and Polarization Measurements at 2.45 GHz for Interference Evaluation between Bluetooth and IEEE 802.11 WLAN," in *Proc. of IEEE APS/URSI Symposium*, Boston, MA, USA, 2001.

[16] R. Karrer, A. Sabharwal, and E. Knightly, "Enabling Large-scale Wireless Broadband: The Case for TAPs," in *Proceedings of HotNets 2003*, Cambridge, MA, November 2003.

[17] Y.-D. Lin and Y.-C. Hsu, "Multihop Cellular: A New Architecture for Wireless Communications," *In Proceedings of IEEE INFOCOM 2000*

[18] Y. Liu and Y. R. Yang, "Reputation propagation and agreement in mobile ad-hoc networks," in *Proceedings of IEEE WCNC 2003*, New Orleans, LA, March 2003.

[19] H. Luo, R. Ramjee, P. Sinha, L. Li and S. Lu, "UCAN: A Unified Cellular and Ad-Hoc Network Architecture," in *Proceedings of Mobicom 2003*, pages 353-367, September, 2003.

[20] P. Papadimitratos, Z.J. Haas, "Secure Routing for Mobile Ad Hoc Networks," *Proceedings of CNDS 2002*, San Antonio, TX, January 27-31, 2002.

[21] C. E. Perkins, *Mobile IP: Design Principles and Practices*, Prentice Hall PTR, ISBN: 0-201-63469-4, 1998

[22] C. E. Perkins and P. Bhagwat, "Highly dynamic destination sequenced distance vector routing (DSDV) for mobile computers," in *ACM SIGCOMM'94*, pp. 234-244, 1994

[23] T. S. Rappaport, *"Wireless Communications: Principles and Practice (2nd Edition),"* Prentice Hall, ISBN: 0130422320, 2002.

[24] K. Roeder, F. Ohrtman, *Wi-Fi Handbook : Building 802.11b Wireless Networks*, McGraw-Hill Professional, April 10, 2003.

[25] P. Santi, D. M. Blough, "The Critical Transmitting Range for Connectivity in Sparse Wireless Ad Hoc Networks," *IEEE Transactions on Mobile Computing*, Vol. 2, No. 1, pp. 25-39, January-March 2003.

[26] T. Small and Z.J. Haas, "The Shared Wireless Infostation Model - A New Ad Hoc Networking Paradigm (or Where there is a Whale, there is a Way)," in *Proceedings of the ACM MobiHoc 2003*, Annapolis, Maryland, June 1-3, 2003

[27] B. Williams, "Creating Public Streets and Pedestrian Connections through the Land Use and Building Permit Process," *Development Services* (503) 823-7004, City of Portland Office of Transportation, July 2002

[28] H. Wu, C. Qios, S. De, and O. Tonguz, "Integrated Cellular and Ad Hoc Relaying Systems: iCAR," *IEEE Journal on Selected Areas in Communications*, 19(10), October 2001.

[29] K. Xu, X. Hong, and Mario Gerla, "Landmark Routing in Ad Hoc Networks with Mobile Backbones," *Journal of Parallel and Distributed Computing (JPDC), Special Issues on Ad Hoc Networks*, 63(2), February 2003

[30] S. Zhong, Y. R. Yang, J. Chen, "Sprite: A Simple, Cheat-Proof, Credit-Based System for Mobile Ad Hoc Networks," *In Proceedings of IEEE INFOCOM'03*, San Francisco, Mar 30 - Apr 3, 2003.

[31] ARC Group, "Mobile Content and Applications," research report, Published: December 2003, http://www.arcgroup.com/

[32] http://www.textually.org/textually/archives/cat_europeanza_sms_pricing_issues.htm

[33] Boingo Wi-Fi Industry White Paper, "Towards Ubiquitous Wireless Broadband," September 2003, http://www.boingo.com/wi-fi_industry_basics.pdf

[34] http://www.cisco.com

[35] *Population density in general:* http://www.world-gazetteer.com/home.htm, http://www.citypopulation.de/, http://www.skyscraperpage.com/, http://www.wikinfo.org/ *Manhattan:* http://en.wikipedia.org/wiki/Manhattan, http://www.greatgridlock.net/NYC/downair.html *Berlin:* http://www.stadtentwicklung.berlin.de/umwelt/umweltatlas/edinh_06.htm

[36] http://internet.hypermart.net

[37] http://www.websiteoptimization.com/bw/0312

[38] http://www.sierrawireless.com/

[39] http://www.nokia.com/phones/nokiad211/

[40] http://www.option.com/products/

[41] http://www.cellular.co.za/stats-europe.htm

[42] http://www.mda-mobiledata.org/, http://www.text.it,

Supporting Real-time Speech on Wireless Ad Hoc Networks: Inter-packet Redundancy, Path Diversity, and Multiple Description Coding

Chi-hsien Lin   Hui Dong   Upamanyu Madhow   Allen Gersho

Department of Electrical and Computer Engineering
University of California – Santa Barbara
Santa Barbara, California
{chlin, huidong, madhow, gersho}@ece.ucsb.edu

ABSTRACT

We consider the problem of supporting real-time traffic over packetized wireless ad hoc networks. Our specific emphasis is on speech, since this is a critical application in many scenarios such as emergency deployment of ad hoc networks. Standard retransmission-based Medium Access Control (MAC) strategies are poorly matched to speech applications, because the payload size for speech as well as for MAC-layer acknowledgements (ACKs) is small compared to the packet header, which contains a large synchronization preamble. In this paper, we show that inter-packet redundancy is significantly more efficient than traditional MAC layer retransmissions, in terms of both network capacity and end-to-end delay. The key observations regarding our design and results are as follows. Because of the small payloads, introducing redundancy across packets only increases the packet transmission time slightly, and hence has negligible impact on the packet collision rate. Thus, we obtain large gains from redundant transmission essentially "for free." Because of the large packet header, elimination of ACKs leads to substantial bandwidth savings. Overall, a combination of inter-packet redundancy (at the MAC layer), path diversity (at the network layer), and multiple description source coding (at the application layer), is shown to provide significant improvements in bandwidth efficiency and delay.

Categories and Subject Descriptors

C.2.1 [**Network Architecture and Design**]: Network communications, Packet-switching networks, Wireless communication

General Terms

Performance, Design

This work was supported by the National Science Foundation under grants ANI-0220118 and EIA-0080134, and by the Office of Naval Research under grant N00014-03-1-0090.

Keywords

Wireless, ad hoc, 802.11, real-time, speech, path diversity

1. INTRODUCTION

Much progress has been made over the past few years on set-up and routing in wireless ad hoc networks [1][2]. Such networks have huge potential for plug-and-play and emergency deployments. A key application in many such scenarios is the support of real-time traffic, especially voice. There are, however, two major technical challenges that must be overcome in order to achieve this. Firstly, ad hoc networks typically have fairly large packet loss rates, due to collisions arising from decentralized transmissions, as well as due to the bit errors inherent to even a collision-free wireless channel. While retransmissions and buffering are effective mechanisms for combating loss for delay-insensitive data applications and non real-time audio, they have limited utility for real-time traffic with strict delay constraints, especially when the traffic must traverse more than one wireless hop. Secondly, the overhead in packetized wireless networks must include not only MAC and network layer information, as in wireline communication, but also a preamble for synchronization, which must typically occur on a packet-by-packet basis. Thus, the efficiency of such networks is poor for small data payloads, as in applications such as voice.

In this paper, we propose and evaluate a number of mechanisms that, compared to simply deploying a speech application over standard ad hoc network infrastructure, provide significant gains in the end-to-end frame loss rate, while simultaneously improving bandwidth efficiency. The main ingredients are as follows:

1) "Free" redundancy and ACK elimination: While the redundancy required for standard forward error control reduces the bandwidth efficiency, we introduce redundancy in a fashion that actually increases bandwidth efficiency, while providing large reductions in end-to-end frame loss rate. This is achieved by exploiting the small payload size of speech, relative to the large headers required in typical packetized wireless networks (e.g., those based on 802.11abg). Specifically, we consider the following simple technique, which can be generalized easily. Instead of sending speech frame d_n, as the payload of the n_{th} packet, we concatenate frames d_{n-1} and d_n as the payload of packet n. Thus, frame n is received successfully if either packet n or packet $n+1$ is successful. On the other hand, putting two speech frames in the payload only slightly increases the overall packet length (which is dominated by the header), and hence the packet

loss probability p is increased only marginally. In addition, the probability of frame loss is reduced by orders of magnitude, since it now scales as p^2, compared to p in a system that does not employ redundancy or retransmissions. Since we do not employ retransmissions, we can eliminate ACKs. A useful point of comparison is a retransmission-based MAC which allows at most one retransmission. Note that the success probability of a frame again scales as p^2, where p is the packet loss rate, but there is significantly more bandwidth expended than in our inter-packet redundancy method. If no retransmission is needed, then we must expend bandwidth in transmitting an ACK (which has roughly the same transmission time as a speech packet, since both have small payloads compared to the header). Further, if a retransmission is needed, again we expend another packet's worth of overhead.

2) Path diversity: Wireless networks suffer from a number of impairments that can lead to a string of consecutive packet losses; these include fading and changes in connectivity caused by mobility. While our inter-packet redundancy method deals well with random packet loss, path diversity is required for handling correlated losses along a given path.

3) Multiple Description Coding: Multiple Description (MD) coding is a source coding techniques that combines well with path diversity. MD coding splits the information sequence (speech, in our case) into two (or more) equally important streams. The received quality is best if both streams are received, but is acceptable even if only one of the streams is received. Thus, sending MD streams over different paths provides robustness against correlated packet losses. However, applying MD will increase the total number of packets that a source node sends into the network. Thus, an effective way to increase bandwidth efficiency is crucial to help release the extra network load, and the inter-packet redundancy scheme is a good candidate.

Each of the preceding schemes produces gains in performance, and typically operates at a different layer of the OSI hierarchy. Inter-packet redundancy, which provides the biggest gains, is ideally deployed at the link/MAC layer, thereby permitting hop-by-hop packet reconstruction for recovering from isolated losses. (It could also be deployed at the application layer, but then hop-by-hop packet reconstruction would not be possible). Path diversity operates at the network layer, while MD coding is at the application layer. We consider an integrated system combining inter-packet redundancy, path diversity, and MD. The system is strongly robust against random packet loss due to the use of inter-packet redundancy, and is resistant to bursty loss because of path diversity and MD. Our simulations show that the proposed system can indeed provide good delay and loss performance for real-time speech over a difficult operating environment such as a multihop wireless network. Given the widespread availability of 802.11 based Wireless Local Area Networks (WLANs) [3][4], the parameters in our performance evaluation are consistent with those of 802.11b, using the Distributed Coordination Function (DCF) for peer-to-peer communication. There are four allowable rates in 802.11b: 1, 2, 5.5 and 11 Mbps. The physical layer convergence protocol (PLCP) handles synchronization using PLCP preamble and header transmitted at 1 Mbps. Thus, if the MAC Protocol Data Unit (MPDU) is small, as for voice applications, and is transmitted at 11 Mbps, for example, then the time to transmit the PLCP preamble and header becomes

significant compared to the transmission of the MPDU, or payload.

The organization of this paper is as follows. Section 2 briefly discusses related work. Section 3 first points out the significant overhead in 802.11, and then describes the proposed inter-packet redundancy scheme and the supplemental packet reconstruction function, then discuss the performance of the proposed scheme. A rough analysis is given, to provide intuition into the expected performance gains (detailed simulations are given in Section 6). An interesting postive feedback in 802.11 MAC is also discussed. Section 4 provides analytical insight into the gains due to path diversity, for a simple Gilbert-Elliott model of correlated packet losses. Section 5 describes the MD coder to be used in our integrated system. Simulation results are provided in Section 6. We first compare the inter-packet redundancy with a conventional retransmission-based MAC. We then demonstrate the additional gain due to path diversity. Finally, the performance of the integrated system is simulated. Section 7 contains concluding remarks.

2. RELATED WORK

Prior attempts to improve the performance of real-time traffic over ad hoc networks include [5-15]. These include ideas such as optimizing packet length, employing forward error control within a packet [5][6], reservation policies [7-10], bandwidth reuse technique [11], retransmission strategy [12], and performance evaluation technique [13]. These papers do not exploit small payloads as we do, and do not employ inter-packet redundancy. Coding across packets has been proposed in the context of wireline networks [14][15]. However, the realization that coding gains can be obtained "for free" for small payloads (relative to the large overhead in packetized wireless communication) appears to be completely novel.

Path diversity has been considered in [16-20], and many multiple path routing protocols based on different routing schemes over wireless ad hoc networks have been proposed [21][22]. However, the focus has mainly been on the support of delay-tolerant data applications, rather than on improving end-to-end performance for real-time traffic. MD coding has a rich history [23-30], and has been discussed for theory [23-26] and image/video applications [27-29]. Applying MD over a wireless environment for both image/video [30][31] and voice [32] is a relatively recent development. To the best of our knowledge, however, there is no prior work that integrates design concepts across multiple layers to provide an integrated system suitable for real-time speech applications over wireless ad hoc networks.

3. INTER-PACKET REDUNDANCY

In the following, we describe our inter-packet redundancy scheme. While the concept is broadly applicable for any packetized system in which the payload is small compared to the overhead, we describe, for concreteness, an example design for supporting voice over an 802.11b based ad hoc network.

3.1 Overhead in a Conventional 802.11 system

First we consider speech transmitted over a conventional system to give a rough idea of the overhead relative to the payload size. Typical speech datagram size ranges from 160 bytes (G.711 at 64 kbps) to 20 bytes (G.729 at 8 kbps) or smaller. These speech datagrams are typically generated every 20 ms. Suppose that neighbors can communicate at the highest rate of 11 Mbps. If we have a 20-byte voice frame to be sent at 11 Mbps, the payload transmission time is

$$8 \times 20 / 11 = 14.55 \ (\mu s),$$

The transmission time of headers, including 802.11 MAC, IP and UDP header is

$$8 \times (28 + 20 + 8) / 11 = 40.73 \ (\mu s),$$

where 28, 20, 8 bytes are 802.11 MAC header, IP header, and UDP header, respectively. The overhead, in terms of transmission time, due to the header is 280%. According to the 802.11 standard, the synchronization time is

$$(144 + 48) / 1 = 192 \ (\mu s),$$

where 144 bits is the length of PLCP preamble, 48 bits is the length of PLCP header, and 1 Mbps is the transmission rate for PLCP preamble and header specified in the 802.11 standard. Thus, the overhead due to the PLCP preamble and header is 1320%. MAC layer ACKs lead to additional overhead, which takes

$$10 + 192 + 8 \times 14 / 11 = 212.18 \ (\mu s),$$

to transmit, where 10 μs is the SIFS, 192 μs is the synchronization time mentioned above, 14 bytes is the size of ACK packets, and 11 Mbps is the transmission rate. Thus, the overhead due to ACK is 1458%. Fig. 1 shows actual proportion of the transmission time devoted to synchronization, headers, and payload. Note that the transmission time for the data and ACK packets are almost the same, and that the overhead relative to the payload is huge in both cases. We do not calculate the overhead introduced by RTS/CTS because according to the 802.11 standard, RTS/CTS mechanism is not used if MAC frame size is smaller or equal to RTS threshold (dot11RtsThreshold) and the default RTS threshold is 3000, which is much larger than small packet sizes that we consider.

The preceding overhead computations were for 802.11b. The percentage overhead is a little smaller, but still very large, for newer OFDM-based WLAN standards such as 802.11a/g. In the latter, the PLCP preamble and header takes 20 μs to transmit, and the MAC, IP, TCP header takes three symbol interval ($3 \times 4 = 12$ μs), while the actual payload only takes one symbol interval (4 μs). The overhead due to synchronization and headers is still very significant (800%). The overhead due to ACK (28 μs) is large as well (700%).

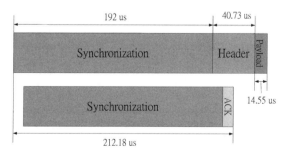

Fig. 1: The overhead of 802.11b

3.2 Adding Redundancy

The inter-packet redundancy scheme that we propose is designed to take advantage of the significant overhead stated above. Consider a voice stream consisting of a sequence of segments $\{d_n\}$. We include a copy of dn in the payload for the packet carrying the next segment d_{n+1}. Thus, d_n is lost only if both the n_{th} and $(n+1)_{th}$ packets are lost. The redundant segment is equivalent to one MAC retransmission, without incurring the significant overhead associated with retransmissions, including the packet overhead incurred in retransmitting a small payload (if the packet is unsuccessful), and the ACK overhead incurred (for indicating a successful packet for positive acknowledgements, as in 802.11, or for indicating failure, as in a negative acknowledgement based scheme). Note that our method can be generalized to provide functionality equivalent to more than one retransmission, by sending more than two consecutive segments in each packet. Whether or not this is useful depends on the reconstruction delay allowed at the intervening nodes (for a MAC layer implementation) or at the receiver (for an application layer implementation). It also depends on how the increase in payload size impacts the overall packet size, and hence the probability of packet loss due to collisions and bit errors. We confine our simulations to the case of two segments per packet, keeping in mind end-to-end delay constraints. In this case, the increase in packet size, and hence in collision probability, is negligible, thus providing the promised "free" coding gains.

Fig. 2 shows a conventional one-retransmission scheme with our inter-packet redundancy scheme. Note that no ACK is transmitted in inter-packet redundancy scheme. Since actual payload is so small compared to synchronization and header transmission time as stated, the increased transmission time due to attaching dn to the $(n+1)_{th}$ packet is negligible.

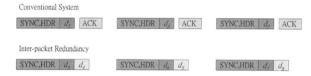

Fig. 2: Conventional system vs. inter-packet redundancy

3.3 Packet Reconstruction

Packet reconstruction is an optional function which can help further reducing the end-to-end segment loss rate. Let (d_n, d_m) denote a packet containing segments d_n and d_m. For simplicity, we assume nodes rearrange the segments in a packet before sending out the packet such that the sequence number of first segment is always larger than the sequence number of the second segment, namely, $n > m$, and $n = m + 1$ if no packet loss has happened.

When a packet (d_n, d_m) arrives at a forwarding node, the node looks into the packet, records segment dn in its memory, and then forwards the packet to its next hop. When another packet from the same stream (d_r, d_s) comes, the node looks into the packet again and checks the sequence number to see if there is any packet got lost. If the difference between larger sequence numbers (the larger sequence number is n for segment (d_n, d_m) and r for segment (d_r, d_s) here) of two successive packets is greater than one, then packet reconstruction is activated. The node picks up d_n from its memory, combines d_n with ds and then sends out a new packet $(d_s,$

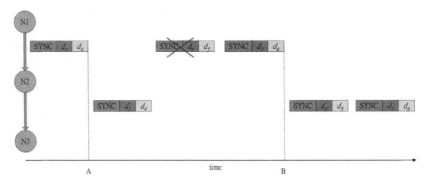

Fig. 3: Packet reconstruction

d_n) before forwarding packet (d_r, d_s). The reconstructed packet is sent only when n is different from s. Note that the d_n in memory will be replaced by dr before (d_r, d_s) is forwarded. The node only keeps d_n in its memory for a predefined period of time, d_n will be deleted from the memory if the node has been waiting too long for the next packet from the same stream.

Fig.3 is an example of packet reconstruction. N1, N2, N3 are forwarding nodes, packets are forwarded form N1 through N2 to N3. At time A, N2 received (d_5, d_4) and put d_5 in its memory. After N2 forwarded (d_5, d_4) to N3, (d_6, d_5) from N1 got lost in the channel. Since N1 did not know that there was a packet loss, it continued sending (d_7, d_6) to N2. At time B, N2 received (d_7, d_6) successfully and detected the loss. It then retrieved d_5 from its memory, reconstructed and sent the lost packet (d_6, d_5) before forwarding (d_7, d_6) to N3. N3 did not even realize that a loss had occurred. Note that the loss recovery occurred without requiring an increase in the number of packets sent, unlike in a conventional retransmission-based scheme.

Packet reconstruction provides a mechanism to recover from isolated packet loss, thus preventing performance degradation when the hop count of a path gets large. With packet reconstruction, only two successive packet losses will cause one segment loss.

3.4 Performance of Inter-packet Redundancy

We want to compare our inter-packet redundancy system with conventional 802.11 MAC. To make the two systems comparable, we focus on an 802.11 system with retransmit limit of 1, namely, the 802.11 MAC has one chance to recover from packet loss, and it requires at least two successive packet losses to incur one segment loss.

In this section, we provide insight into the relative performance of the two methods via rough calculations. Their detailed behavior and performance is investigated using simulations in Section 6.

3.4.1 Packet Error Rate and Collision Probability

First we consider the probability of single transmission failure, namely, packet error rate and collision probability. The impact of retransmissions is discussed later.

We show that the increase in payload size of inter-packet redundancy slightly increases the packet loss rate due to bit errors. However, the decrease in overall traffic due to elimination of ACKs and retransmissions results in a significant decrease in the

packet loss rate due to collisions. At the operating Signal-to-Noise Ratios of interest, bit errors are infrequent, so that the second effect is much more significant. Thus, the inter-packet redundancy technique provides significant gains over the one-retransmit scheme.

Packets can be dropped at the receiver end if some bit errors has occurred during the transmission. Here we use packet error rate (PER) to compare the two systems. Denote PER of the original packet as P_p, and the PER of the packet with redundancy as P_p'. Assuming the two systems have the same bit error rate and bit errors are independent, we can estimate the relationship between P_p and P_p':

$$P_p = 1 - (1 - P_p')^{\frac{b}{a}} = 1 - (1 - \frac{b}{a}P_p' + h.o.t) \approx \frac{b}{a}P_p',$$

where a is the number of bits in the packet with redundancy and b is the number of bits in the original packet. Higher order terms are negligible for small P_p'. Note that a, b both include MAC, IP, and UDP headers. For the example in Section 3.1, $a = 96$ and $b = 76$, so $\frac{b}{a}$ is 0.79, which is also the PER ratio of inter-packet redundancy system to one-retransmission 802.11 MAC.

We now consider the collision probability. Since RTS/CTS is not employed for small packets that we are considering as stated before, these short packets can only rely on CSMA/CA to avoid collision, and are therefore vulnerable to the hidden terminal problem.

For hidden nodes, the collision probability is roughly proportional to transmission time per packet and the average channel busy time. Note that the transmission time here includes and is dominated by synchronization time. According to parameters in Section 3.1, the packet transmission time ratio of one-retransmission 802.11 MAC to inter-packet redundancy is

$$\frac{192+40.73+14.55}{192+40.73+14.55\times2} = \frac{247.28}{261.83} = 0.94,$$

and the average channel busy time ratio of one-retransmission 802.11 MAC to inter-packet redundancy is

$$\frac{247.28+212.18}{261.83} = \frac{460.45}{261.83} = 1.75,$$

where 212.18 is the time used by ACK. The use of ACK packets makes the channel busy time much longer, and therefore degrades the performance of the one-retransmission scheme.

For non-hidden nodes, collision probability is not so closely tied to the packet transmission time. Instead, it is determined by the collision avoidance mechanism. From this point of view, having more packets to contend for the channel increases the collision probability. The main difference between the two systems here is that inter-packet redundancy has no ACK, backoff and retransmission. For one-retransmission 802.11 MAC, backoff can reduce the probability of collision, but this only helps the retransmissions. While retransmitted packets are less likely to incur collisions due to its larger contension window, they still increase the number of packets contending for the channel relative to the inter-packet redundancy scheme. Further, while ACKs do not suffer collisions with packets from non-hidden nodes in the 802.11 MAC, the cost in terms of transmission time (and hence overall capacity) is significant.

3.4.2 Behavior under High Network Loads
Now we take the retransmission into account and consider the behavior of the two systems when the network load is heavy.

Fig. 4 shows the transmission cycle of the one-retransmission 802.11 MAC. There is an interesting postive feedback in 802.11 MAC: When the network load is heavy, the collision probability gets high, and thus more losses incur more retransmission. However, this behavior implies that nodes are always injecting more traffic into the network when network condition is bad and making the condition worse. On the other hand, more retransmission implies less channel efficiency, so 802.11 MAC is always spending more transmission chances to transmit a packet when the chances are rare. Another important factor is that, even after the packet is successfully received, retransmission can still happen because there is still a chance that the ACK packet get lost and cannot be received by the sender. Besides, no matter how many retransmissions is allowed, the last ACK is always useless since the sender cannot retransmit due to the limit. These factors make the performance of 802.11 MAC degrade dramatically when traffic load gets high.

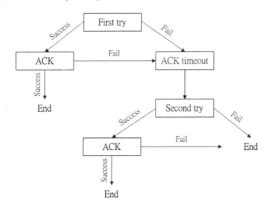

**Fig. 4: Transmission cycle of
one-retransmission 802.11 MAC**

For our inter-packet redundancy technique, things are totally different. Although we also have the ability to recover form single packet loss, we are not introducing any unexpected traffic at any time. This is especially good for real-time applications since we can keep end-to-end delay under control when traffic load gets heavier. Also, since increased traffic does not imply more retransmissions, the loss rate increases gradually with the traffic

load rather than the steep increase seen in retransmission-based schemes once the traffic load exceeds a threshold.

4. PATH DIVERSITY
Packet losses in ad hoc networks can be categorized into two kinds: random losses and bursty losses. Bursty losses can be due to routing failure, bursty noise, IP buffer overflow, and mobility. Random losses usually result from collisions or noise. Both types of packet loss are much more prevalent in wireless networks than in wired networks. Bursty losses are a more severe impairment for real-time applications such as speech, since interpolation-based error resilience techniques can be employed at the destination to alleviate degradation due to random loss. While coding and interleaving over a large number of packets is one possible mechanism for dealing with bursty loss, it does not apply to delay-constrained applications such as speech. In this situation, path diversity is a simple but effective method, assuming that we can find multiple paths which are unlikely to incur a burst of packet losses at the same time (e.g, use the Disjoint Path Selection Protocol in [22]). In this section, we provide a quick calculation for estimating performance gains from path diversity, using a Gilbert-Elliott model for bursty packet losses. Detailed simulations are postponed to Section 4.

Suppose there are two paths, path 1 and path 2, modeled by two independent Gilbert-Elliott models, each as shown in Fig. 5. Paths can be in either good or bad state, and can switch their states according to Markov model, depicted in Fig. 5. Here the transition rates from good state to bad state and from bad state to good state are λ and μ, respectively. Packet loss rates of a path are different while in different states, and packet loss rates of different paths in the same state are not necessarily the same. Here we set the packet loss rate in bad state to be 1 for both paths, and set the packet loss rate in good state to be p_1 and p_2 for path 1 and path 2, respectively.

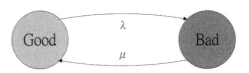

Fig. 5: Two state Gilbert-Elliott model

If only one out of the two paths is randomly chosen to transmit through, the expected loss rate is $\frac{\mu\left(\frac{p_1+p_2}{2}\right)+\lambda}{\lambda+\mu}$, and the expected time in bad state, i.e., the expected burst time, is $\frac{1}{\mu}$.

If we use two paths at the same time, the Gilbert-Elliott model leads to a two dimensional Markov chain as shown in Fig. 6. There are four states, (G, G), (G, B), (B, G), and (B, B), where the first element represents the state of path 1 (G for good, B for Bad) and the second element represents the state of path 2. Since the two paths are assumed independent, the transition rates are still λ and μ as in Fig. 5.

When both path 1 and path 2 are used, the expected loss rate remains the same as randomly choosing a path, which is $\frac{\mu\left(\frac{p_1+p_2}{2}\right)+\lambda}{\lambda+\mu}$. But the average burst time (the average time stay in (B, B) state) is halved, which is $\frac{1}{2\mu}$. Of course, when one path is significantly

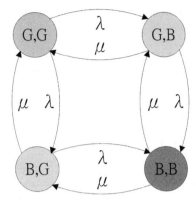

Fig. 6: Gilbert-Elliott model for two paths

superior to another (e.g. in terms of packet loss rate in Good state), and the identity of the better path is known to the source (this information is typically not available from standard ad hoc routing protocols), then the tradeoffs between random and burst loss, from an end-to-end application perspective, must be carefully considered in deciding whether, and in what form, to use path diversity. However, these simple calculations do provide motivation for devising application layer strategies that can exploit path diversity, which leads us into the MD coding techniques discussed next.

5. MULTIPLE DESCRIPTION CODING

An MD coder produces two or more descriptions, or coded bit streams, from a given source signal. Fig. 7 shows the block diagram of the coder and decoder for two descriptions. Bit streams independently represent "coarse" descriptions of the source (output 1 and output 2), while multiple descriptions jointly convey a "refined" source representation (output 0). Different bit streams are transmitted to the receiver separately, usually through different paths to profit from path diversity as showed in last section. If any one of the bit streams is received, the decoder can choose a proper decoding procedure and provide a degraded but acceptable quality of speech. If all the bit streams are successfully received, the decoder will be able to reconstruct better speech. Note that all the descriptions in a MD code are equally important unlike layered coding in which the higher layers are useless if the coarsest layer is not received correctly.

One way to design MD speech coding is to allow some redundancy of two descriptions. So as long as at least one description is received, the decoder will have the basic information and be able to estimate the lost information, which enables the decoder to reconstruct a degraded but acceptable quality.

The MD coder used in this paper is designed to make use of the wideband speech standard, AMR-WB. Each frame of AMR-WB

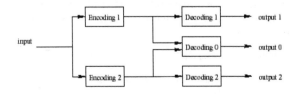

Fig. 7: Block diagram of multiple description coder

contains 20 ms speech data, and the size of the 12.65 kbps frame is 253 bits. The MD coder splits the bit-streams into two redundant sub-streams, 136 bits and 134 bits, by directly selecting overlapping subsets of encoded data generated for each frame. A full quality at 12.65 kbps is ensured if two descriptions are received and a degraded quality at 6.8 kbps is achieved if only one description is received.

A very important thing to be noticed about MD is that, when MD splits the source in to two bit streams, the inter frame interval still remains the same, in other words, the total number of packets that a source node sends into the network is actually doubled. If everyone on the network adopts path diversity and MD coding, the number of connections that a network can support will be reduced given a fixed network capacity. Thus, it is very important to combine MD together with the inter-packet redundancy since it halves the number of packets transmitted into the network and provides better bandwidth efficiency.

6. PERFORMANCE EVALUATION

Glomosim, a scalable simulation library for wireless network systems developed by UCLA, is used as the simulation tool for evaluating the proposed techniques [33]. For the simulation, 16 nodes are uniformly distributed in an area of 800 × 800 square meters. The distributed coordination function (DCF) of IEEE 802.11 is used as the MAC layer, and a two-ray propagation model is used to model signal propagation. The transmission rate is chosen to be 11Mbps.The traffic pattern of our tagged stream is selected as CBR. At each node, the background traffic is modeled as Poisson, with the destinations chosen among its nearest neighbors for simplifying the simulation model. Each simulation is executed for 5 minutes of simulation time and each speech stream has 13000 frames with a 20 ms inter-frame time. All IR systems in the following simulations include the packet reconstruction function.

In this section, IR is first compared with the retransmission-based scheme (both restricted to one retransmission and the standard). Then, using path diversity, MD coding with two paths and single path transmission without MD coding are compared. Finally, IR, path diversity and MD coding are integrated into a single system for performance evaluation. End-to-end performance measures are considered, including segment loss rate, delay, and burst size for lost segments (number of consecutive segments lost).

6.1 IR vs. Retransmission-based MAC

To compare the proposed IR scheme with a conventional system fairly, 802.11 MAC's retransmission limit is first restricted to 1, and then restored to 7, which is specified in the standard for short packets. In Glomosim, a priority queue with size 100 packets is implemented at the IP layer of each node, providing control packets, real-time data, and non real-time data different priorities to get into MAC layer. Both real-time and non real-time background traffic are simulated to examine the impact of their priorities. The size of non-real-time background traffic packets is 1500 bytes and the size of real-time background traffic packets is 20 bytes. The segment size of the tagged voice stream is 20 bytes, excluding all headers. All short packets, both background and tagged voice packets, in one simulation use the same type of MAC, namely, IR or retransmission-based MAC. Long packets, like the non-real-time background packets, use retransmission-based MAC. For the IR scheme, a segment is considered lost if all redundant copies, both

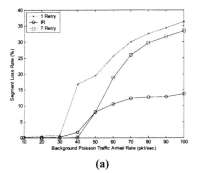 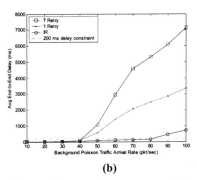 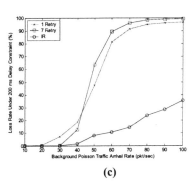

(a)   (b)   (c)

Fig. 8: (a) Segment loss rate, (b)average end-to-end delay, and (c) segment loss rate under 200 ms delay constraint   for non real-time background traffic

original and reconstructed, of that segment are lost. The end-to-end delay is calculated as the difference between the arrival time of the first successful copy and the transmit time of the first copy. For a desired 200 ms delay constraint, the loss rate is also calculatedfor which all packets with end-to-end delay larger than 200 ms are considered lost.

Fig. 8(a) shows the segment loss rate with non-real-time background traffic. Results of the one-retransmission (1-retry), the IR, and the standard (7-retry) systems are shown. It is found that the segment loss rate of the 1-retry system grows much faster than the IR system, especially when the background traffic arrival rate ranges from 30 to 50 packets per second. This improvement of the IR system over the 1-retry system comes from the different behavior of the two systems under heavy traffic. When transmission opportunities become rarer due to the higher load, the IR system still transmits a new segment together with an old segment at every opportunity. In contrast, the 1-retry scheme usually needs two opportunities to send each segment, increasing the traffic in the network, thus increasing both delay (in waiting for transmission opportunities) and loss rate (due to higher traffic loads). On the other hand, since the 7-retry system has seven chances to recover from transmission failures, the segment loss rate is zero when load is light. But its performance degrades after the background traffic arrival rate exceeds 40 packets per second. At high traffic load, IR performs better than the two retransmission based systems.

Fig. 8(b) shows the average end-to-end delay under non-real-time background traffic. When the background traffic arrival rate is greater than 40 packets per second, the average end-to-end delay of

retransmission-based systems grows fast and exceeds the desired 200 ms delay, while the average end-to-end delay of the IR system still remains less than the delay constraint until 80 packets per second. This is not only due to the effective reduction of transmission time in the IR scheme, but also its inherently stable delay characteristic. Note that the average end-to-end delay of the 7-retry system is longer than the 1-retry system when the traffic load is heavy due to too many retransmissions.

Fig. 8(c) displays the segment loss rate for a 200 ms delay constraint. The figure shows the actual loss rate that the receiver end-user sees because voice packets violating the delay constraint are discarded at the receiver. According to Fig. 8(a), the segment loss rate of the 7-retry system is less than 35% when the background traffic arrival rate is 100 packets per second, but it becomes 94.42% when delay constraint is applied. In other words, almost all packets in the 7-retry system are too delayed to be useful when they arrive at the receiver. On the other hand, having a delay constraint does not affect the IR system so much.

Fig. 9(a) shows the segment loss rate for different arrival rates of real-time background traffic. Here too, the performance of the IR scheme has significant gains over the retransmission based systems. In this figure, it is shown that having more retransmissions significantly degrades the system performance. And the difference between the performances here is more significant than Fig.8(a). The key factor of this effect is the priority of background traffic packets. Since the IP queue size is not infinite, packets will be dropped when the queue gets full. And since the voice packets have priority over the non-real-time

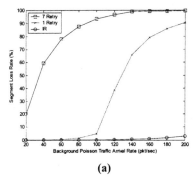 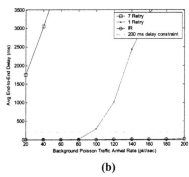 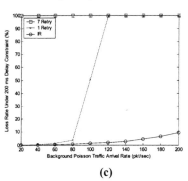

(a)   (b)   (c)

Fig. 9: (a) Segment loss rate, (b)average end-to-end delay, and (c) segment loss rate under 200 ms delay constraint   for real-time background traffic

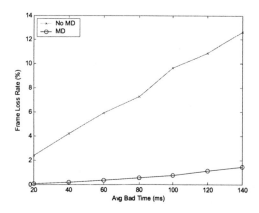

Fig. 10: Frame loss rate for MD and non-MD systems

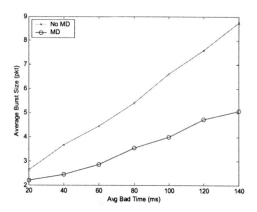

Fig. 11: Average burst size for MD and non-MD systems

background packets in Figs. 8(a), 8(b), and 8(c), the low priority background packets will be dropped first while the voice packets are not likely to be dropped due to IP buffer overflow, even when the packet arrival rate is higher than the system service capability. However, if the background traffic has the same priority as the tagged voice packets, every packet is equally likely to be dropped. Thus, the segment loss rates here are much higher than in Fig. 8(a). Since the IR system has much better bandwidth efficiency, the system can deal with heavier background traffic and does not suffer from IP buffer overflow. The real-time background traffic can be viewed as a coarse model for a system with many real-time connections (not simulated because of the complexity of tracking so many connections), even though only the performance seen by the tagged speech connection is monitored. With this interpretation, Fig. 9(a) also implies that IR can significantly increase the network capacity by allowing more connections in the network for a given segment loss rate.

Fig. 9(b) displays the average end-to-end delay while using short real-time packets as background traffic. While the average end-to-end delay of retransmission-based systems start to increase steeply, the inter-packet redundancy system keeps its end-to-end delay small because it is not affected by IP buffer overflow. Fig. 9(c) shows the segment loss rate after applying the delay constraint. The loss rate of retransmission-based systems becomes extremely large once the postive feedback described in Section 3.4 becomes significant.

6.2 Two Paths with MD vs.Single Path without MD

Here, simulation results are used to show the gains achieved via path diversity. A Gilbert-Elliott model is added into each path in order to model bursty losses, without incurring the complexity of modeling specific causes of burst loss such as fading and mobility. When the state of a path is bad, all the packets are lost; when the state is good, the packets can still be lost due to collision or noise. The average dwell time in the good state is set to 1000 ms. The background traffic is comprised of 20-byte real-time packets with an arrival rate of 30 packets per second. The average time in the bad state ("bad time") is varied from 20 to 140 ms to simulate the impact of channel burstiness. As in Section 5, the frame size for each description in MD coding is 17 bytes and the payload size becomes 23 bytes after adding the sequence number and

timestamp to each frame. The payload size for a non-MD coding system is 38 bytes. The two MD bit streams are transmitted through different paths, while the non-MD coded system only uses one path. Both systems adopt the one-retransmission 802.11 MAC.

Fig. 10 shows the frame loss rate of both the MD and non-MD systems. For the MD system, a frame loss occurs only when both descriptions are lost. The frame loss rate of the single path system is 11.08 times greater than the two path MD system when the average bad time is 20 ms, and is 8.57 times greater when the average bad time is 140 ms. This performance measure does not account for the fact that, when one of the two MD streams is received, the speech quality is not as good as when a single description code is reliably received. This is consistent with the objective of providing acceptable quality in a difficult environment.

Fig. 11 shows the average size of bursts of end-to-end lost segments for the two systems. Only successive losses are considered as contributing to a burst, so that the smallest possible burst size is 2. When the channels are not bursty (average bad state dwell time of 20 ms), random loss is the dominant source of loss, hence both systems have average burst sizes slightly larger than 2. But when the channels get bursty, the average burst size of the two path system is approaching half the average burst size of the single path system. The average end-to-end delay is not shown here, since the background traffic load, which is the dominant factor of end-to-end delay, is not changed.

6.3 Integrated System vs. Single Path + 1-Retry

The performance of the integrated system, which combines inter-packet redundancy, path diversity and multiple description coding, is evaluated in this subsection. The integrated system is only compared with the single path system with one-retransmission MAC since the one-retransmission MAC actually performs better than the standard seven-retransmission system as shown in Section 6.1. Here, the average bad time is fixed to 100 ms, and the arrival rate of the real-time background traffic is varied. The packet size of the background traffic is still 20 bytes.

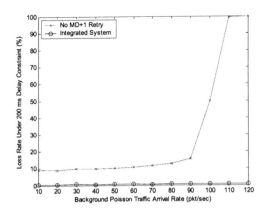

Fig. 12: Segment loss rate for the integrated system and retransmission-based system under 200 ms delay constraint

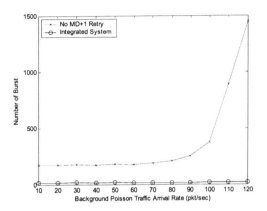

Fig 13: Number of error bursts for the integrated system and retransmission-based system

Fig. 12 shows the frame loss rate after applying the delay constraint. The integrated system has a consistently low frame loss rate (lower than 0.8%). On the other hand, the loss rate for the single path system ranges from 8 ~ 14% as the background traffic arrival rate grows from 10 to 80 packets per second. The loss rate grows dramatically once the background traffic arrival rate becomes greater than 90 packets per second. Fig. 13 shows the number of end-to-end loss bursts during the whole session for these two systems. The burst number of the retransmission-based system remains around 10 times the burst number of the integrated system before the background traffic arrival rate reaches 100 packets per second. Fig. 12 and Fig. 13 show the proposed integrated system employing IR is superior both in terms of loss bursts and loss rates.

7. CONCLUSION

The conventional wisdom regarding packetized systems is that one would not employ inter-packet coding above the physical layer, unless there is a delay constraint. This is because retransmission-based recovery is always considered superior to adding redundancy on an erasures channel. In this paper, we show that this is simply not true for packetized systems in which the payload is small compared to the overhead, as is the case for the real-time speech application considered here. The large overhead makes both retransmissions and ACKs expensive in terms of the additional network load generated. In this case, proactive use of inter-packet redundancy provides large coding gains while incurring negligible bandwidth penalties, thus providing better performance both in terms of capacity and delay compared to retransmission-based error recovery. Our integrated system includes the additional mechanisms of path diversity and MD coding. Path diversity alleviates bursty losses, and is particularly useful for delay-sensitive applications in which data cannot be interleaved with a large enough delay to convert burst losses into random loss. MD source coding is a natural means of exploiting path diversity, by sending equally important streams in parallel over multiple paths. Overall, the simulated performance shows that it is indeed possible to support speech with reasonable end-to-end quality over multihop wireless networks with both random and bursty loss, and with both real-time and non real-time traffic. Note that we do not require complex bandwidth estimation and

reservation methods, and require only that individual node schedulers give priority to real-time traffic over non real-time traffic.

An important topic for future work is to demonstrate working systems that incorporate the strategies presented in this paper. It is also important to characterize the practical capacity regions for ad hoc networks with both real-time and non real-time traffic under such optimized strategies. Theoretical rules of thumb would be very useful in this setting, since purely simulation-based approaches suffer from problems of scaling to a large number of connections (since the bandwidth required by an individual speech connection is much smaller than the link bandwidth in a typical 802.11 type network, it is difficult to simulate a large enough number of connections to determine the maximum number the network can support). The effect of non real-time traffic on real-time QoS is particularly important to understand: even when node schedulers give priority to real-time traffic, non real-time traffic from other nodes, including hidden nodes, can cause interference.

8. ACKNOWLEDGMENTS

The authors would like to thank Y. C. Chen for helpful discussions, Dr. E. M. Belding-Royer for her valuable suggestions and Dr. J. D. Gibson for his careful review of the paper.

9. REFERENCES

[1] S. Corson and J. Macker, "Mobile ad hoc networking (MANET): Routing protocol performance issues and evaluation consideration," RFC 2501, Jan. 1999.

[2] E.M. Royer, C.-K. Toh,"A review of current routing protocols for ad hoc mobile wireless networks," *IEEE Personal Communications*, vol. 6, pp. 46 -55, April 1999

[3] IEEE standard for Wireless LAN Medium Access Control (MAC) and Physical Layer (PHY) specifications, ISO/IEC 8802-11:1999(E), Aug. 1999.

[4] http://www.ieee802.org/11.

[5] M.G. Arranz, R. Aguero, L. Murioz, P. Mahonen, "Behavior of UDP-Based Application over IEEE 802.11 Wireless Networks," *12th IEEE International Symposium on Personal,*

Indoor and Mobile Radio Communications, vol. 2 , pp. F-72 -F-77, 30 Sept.-3 Oct. 2001.

[6] L. Munoz, M. Garcia, J. Choque, R. Aguero, P. Mahonen, "Optimizing Internet flows over IEEE 802.11b wireless local area networks: a performance-enhancing proxy based on forward error correction," *IEEE Communications Magazine,* vol. 39, pp. 60 -67, Dec. 2001.

[7] M. I. Kazantzidis, L. Wang, and M. Gerla, "On fairness and efficiency of adaptive audio application layers for multihop wireless networks," *IEEE International Workshop on Mobile Multimedia Communications,* pp. 357–362, Nov. 1999.

[8] V. N. Muthiah and W. C. Wong, "A speech-optimised multiple access scheme for a mobile ad hoc network," *1st Annual Workshop on Mobile and Ad Hoc Networking and Computing,* pp. 127–128, Aug. 2000.

[9] I. Joe and S. G. Batsell, "Reservation CSMA/CA for multimedia traffic over mobile ad hoc networks," *IEEE International Conference on Communication,* vol. 3, pp. 1714–1718, 2000.

[10] A. Servetti and J. C. De Martin, "Adaptive interactive speech transmission over 802.11 wireless LANs", *Proc. IEEE International Workshop on DSP in mobile and Vehicular Systems,* Nagoya, Japan, April 2003.

[11] C.-H.R. Lin and M. Gerla, "A distributed control scheme in multi-hop packet radio networks for voice/data traffic support," *IEEE International Conference on Communication,* vol. 2, pp. 1238–1242, 1995.

[12] S. Aramvith, Chia-Wen Lin, S. Roy, and Ming-Ting Sun, "Wireless Video Transport Using Conditional Retransmission and Low-Delay Interleaving," *IEEE Trans. on Circuits and Systems for Video Technology,* vol. 12, No. 6, Jun 2002.

[13] H. Wu, C. Hung, M. Gerla, and R. Bagrodia, "Speech support in wireless, multihop networks," *3rd International Symposium on Paprallel Architectures, Algorithms, and Networks Proc.,* pp. 282–288, Dec. 1997.

[14] S.-Y.R. Li, R.W. Yeung, C. Ning, " Linear network coding," *IEEE Trans. on Information Theory,* vol. 49, pp. 371 -381, Feb. 2003.

[15] R. Koetter, M. Medard, " An algebraic approach to network coding," *IEEE/ACM Trans. on Networking,* vol. 11, Oct. 2003.

[16] N. F. Maxemchuck, "Dispersity routing in store and forward networks," *Ph.D. dissertation, Univ. Pennsylvania, Philadelphia,* May 1975.

[17] T. T. Lee and S. C. Liew, "Parallel communications for ATM network control and management," *Proc. GLOBECOM'93,* Nov. 1993, pp. 442–446.

[18] N. T. Plotkin and P. P. Varaiya, "Performance analysis of parallel atm connections for gigabit speed applications," *Proc. INFOCOM'93,* pp. 1186–1193.

[19] E. Ayanoglu, I. Chih-Lin, R. Gitlin, and J. Mazo, "Diversity coding for self-healing and fault tolerant communication

networks," *IEEE Trans. on Communications,* vol. COM-41, pp. 1677–1688, Nov. 1993.

[20] R. Krishnan and J. A. Silvester, "Choice of allocation granularity in multipath source routing schemes," *Proc. INFOCOM'93,* Mar. 1993, pp. 322–329.

[21] Mahesh K. Marina, Samir R. Das, "On-demand multipath distance vector routing in ad hoc networks," *9th International Conference on Network Protocols,* 11-14 Nov. 2001

[22] Panagiotis Papadimitratos, Zygmunt J. Haas, Emin Gun Sirer, "Path set selection in mobile ad hoc networks," *Proc. 3rd ACM international symposium on Mobile ad hoc networking & computing,* Jun. 2002

[23] V. A. Vaishampayan, "Design of multiple description scalar quantizer," *IEEE Trans. Inform. Theory,* vol. 39, pp. 821–834, May 1993.

[24] Y. Wang, M. Orchard, V. Vaishampayan, and A. Reibman, "Multiple description coding using pairwise correlating transforms," *IEEE Trans. Image Processing,* vol. 10, pp. 351–366, Mar. 2001.

[25] V. K. Goyal and J. Kovacevic, "Generalized multiple description coding with correlating transforms," *IEEE Trans. Inform. Theory,* vol. 47, pp. 2199–2224, Sept. 2001.

[26] H. Jafarkhani and V. Tarokh, "Multiple description trellis coded quantization," *IEEE Trans. Communications,* vol. 47, pp. 799–803, June 1999.

[27] D. Chung and Y.Wang, "Multiple description image coding using signal decomposition and reconstruction based on lapped orthogonal transforms," *IEEE Trans. Circuits Syst. Video Technol.,* vol. 9, pp. 895–908, Sept. 1999.

[28] S. D. Servetto, K. Ramchandran, V. Vaishampayan, and K. Nahrstedt, "Multiple description wavelet based image coding," *Proc. ICIP'98,* pp. 659–663.

[29] J. G. Apostolopoulos, T. Wong, W. Tan, S. Wee, "On Multiple Description Streaming with Content Delivery Networks," *IEEE INFOCOM,* June 2002

[30] Nitin Gogate, Doo-Man Chung, Shivendra S. Panwar, Yao Wang, "Supporting image and video applications in a multihop radio environment using path eiversity and multiple description coding," *IEEE Jounral on Selected Areas in Communications,* vol. 12, pp. 777 -792, Sep 2002.

[31] A. Miu, J. G. Apostolopoulos, W. Tan, M. Trott, "Low-Latency Wireless Video Over 802.11 Networks Using Path Diversity," *Proc. of the IEEE International Conference on Multimedia and Expo(ICME),* Baltimore, MD, July, 2003

[32] H. Dong, A. Gersho, J. Gibson, and V. Cuperman, "A Multiple Description Speech Coder Based on AMR-WB for Mobile Ad Hoc Networks," *IEEE International Conference of Acoustics, Speech, and Signal Processing, Montreal,* Canada, May 17-21, 2004

[33] X. Zeng, R. Bagrodia, M. Gerla, "GloMoSim: a library for parallel simulation of large-scale wireless networks," *Proc. PADS 98,* 26-29 May 1998, pp. 154–161

Stateful Publish-Subscribe for Mobile Environments

Mihail Ionescu
Computer Science Department, Rutgers University
110 Frelinghuysen RD, Piscataway
NJ, 08854-8058, USA
1-732-445-3999

mihaii@caip.rutgers.edu

Ivan Marsic
Center for Advanced Information Processing, Rutgers University
96 Frelinghuysen RD, Piscataway
NJ, 08854-8058, USA
1-732-445-6399

marsic@caip.rutgers.edu

ABSTRACT

The Publish-Subscribe paradigm has become an important architectural style for designing distributed systems. In the recent years, we have been witnessing an increasing demand for supporting publish-subscribe for mobile systems in wireless environments. In this paper we present SUBLIM, a stateful model for publish-subscribe systems, which is suitable for mobile systems. In our system, the server maintains a state for each client, which contains variables that describe the properties of particular clients, such as the quality of the connection or the battery utilization. The interest of each subscriber can be expressed in terms of these variables. Based on the subscriber interests, an associated agent is created on the server. The agent filters the data that reach the subscriber based on the content of the message and the current subscriber state. Experimental results show good performance and scalability of our approach.

Categories and Subject Descriptors

C.2.1 [**Network Architecture and Design**]: Wireless communication. D.4.4. [**Communications Management**]: Network communication.

General Terms: Performance, Design.

Keywords: Stateful publish-subscribe, run-time compiling, sublim.

1. INTRODUCTION

Publish-subscribe is now a well-known paradigm for building robust, large-scale distributed systems. In this architecture, the components interact via publishing messages and subscribing to classes of messages through a centralized component (the server). The set of conditions imposed by a subscriber define the subscriber's policy. A common characteristic of these architectures is that the server does not maintain any state about the subscribers. The server keeps only the address of the subscriber (usually an IP address) and the associated policy. The policy is usually specified in a language based on first-order predicates and decides whether the

message is sent to the particular subscriber only based on the content of the message.

Publish-subscribe promises to be a very good paradigm for developing applications for mobile environments due to the loose coupling of the components that are involved. However, the limited and highly dynamic resources of the mobile devices (such as network connection, energy supply) impose major challenges on the way the subscribers should specify their policies. The proliferation of Wi-Fi wireless LANs makes this situation even worse, since the clients need to be able to make a smooth transition from the hotspots to areas with normal or poor coverage. The main question we address in this paper is: What type of policies do the subscribers need to specify in order to be able to receive the right information in such environments?

There has been a considerable body of research in the last years on supporting publish-subscribe in mobile environments. The main focus was to improve the performance of the overall system in the presence of sporadic connectivity, using different algorithms for finding the best path to the mobile subscriber or support for offline work. In this paper we argue that we need novel mechanisms for the deployment of the policies themselves in order to support publish-subscribe in mobile environments.

A wide range of applications use the publish-subscribe paradigm, from simple stock quotes to civil disaster recovery or customized enterprise applications. To illustrate the generality of the problems that are faced in mobile environments we will give short examples in each category.

Suppose that in a stock market application, a client wants to receive information about the companies DELL or IBM when the stock quote drops under $30. The information about the company may contain additional data, like company name, percentage change and the number of shares that were transacted in a given period. In a mobile environment, the client would like to specify how the conditions change depending on information like the quality of the connection or battery power left. For example, the client may be interested in getting the stock quotes of IBM, DELL and Microsoft if it has a good connection and a lot of battery power left, but only in IBM if the price goes under $20 and less than 30% battery power is available. Also, if the client has a bad connection, it might be interested in getting only the actual quote for the stock price with no additional information like percentage change and the number of shares that were transacted.

Consider now a civil disaster recovery scenario, where agents in the field are equipped with mobile devices and are searching in a specified region. They can receive periodical updates from different data sources about any activity that was discovered in the area. The information can be a complex map or just some simple map indications. It should be possible for the agent to define its policy based on the current state it is in, depending on the network connectivity, remaining power, etc.

Finally, let us consider a large hospital with an Intensive Care Unit. The patients are monitored continuously using specialized devices. Ideally, the doctor would like to receive all of the data for the patients he is responsible for, if he has a good network connection and the battery of his PDA is charged. However, if the network connection deteriorates, he would like to be sure that he receives the most important data (such as the pulse) and maybe compressed versions of the least important data, even with loss of information such as the electrocardiogram. Moreover, if the battery power is very low, he would like only a summary of the state of the patients, at a regular interval of time.

We believe that the main characteristics that make the publish-subscribe paradigm suitable for mobile systems are the following:

- Stateful subscribers

  The server must maintain a state associated with each subscriber. The state is an abstract concept and can be composed of any number of variables that describe the properties of a particular client.

- Rich language for expressing the conditions

  A first-order predicate-based language which allows only interaction with the content of the message is no longer suitable for mobile systems. The language in which the conditions are written should be able to allow the interaction with the content of the message and with the state of the subscriber. It should also be able to modify the message content that is actually sent to the client according to the policy.

In this paper we present the SUBLIM system that supports publish-subscribe for mobile systems and meets the above requirements. Our approach can be used in conjunction with current systems, where the server maintains states only for some of the subscribers facilitating an incremental deployment. In our system, the servers are running on the access points of the wireless environments, ensuring a good scalability by supporting a potentially large number of subscribers with minimal overhead.

The paper is organized as follows. We first present the main contributions of this work and review related work in this area. Next, we describe our approach for supporting publish-subscribe in mobile systems. We then present a detailed example, elaborating the process of creation and deployment of the user policies. System performance is evaluated using this example application. Finally, we discuss further work and conclude the paper.

2. OVERVIEW AND MAJOR CONTRIBUTIONS

In our architecture, the subscribers are using wireless devices, such as PDAs. The server maintains a state for each client which is a tuple-space and can be composed of any number of variables that describe the properties of the subscriber. Examples of such variables are the quality of the connection, the battery utilization or the device size. The policy of each subscriber will be able to filter the messages that reach the subscriber based on the content of the message and

the subscriber state. We propose a domain specific language that will allow for supporting arbitrarily complex policies in a secure and efficient manner.

There are two major contributions of this paper. First, we introduce the concept of a stateful publish-subscribe system which is suitable for mobile systems, and present a new approach for specifying the subscribers policies. Also, our approach can be used in conjunction with the existing approaches for publish-subscribe systems in which server maintains the state for some of clients. Second, we present an efficient implementation of a stateful publish-subscribe system and provide a relevant example application.

3. RELATED WORK

There are two main types of publish-subscribe systems, as described in literature. Content-based publish-subscribe systems are intended for content distribution over a communication network. The most important content-based publish-subscribe systems that we are aware of are GRYPHON [1], SIENA [8], ELVIN [14] and KERYX [12]. In such systems, the subscription criteria filters messages based solely on their content.

The subscription languages of the existing systems are similar to each other and are usually based on the first-order predicates. The filtering algorithm in these situations can be made very efficient using techniques such as parallel search trees [1] or binary decision diagrams [2]. As a general observation, a trade-off between expressive subscription languages and highly efficient filtering engines is characteristic to these systems.

A couple of publish-subscribe systems were proposed to explicitly support the mobile clients in wireless environments. The JEDI system [5] provides a set of dispatching servers, organized in a tree to simplify the message routing. A client that joins the system is free of choosing one of the dispatching servers, and then using it to subscribe and receive event notifications. To support mobility, JEDI supports disconnection and reconnection of clients to the dispatching system by introducing two functions: move-out and move-in. More recent updates of the system [6,7] add several features such as managing link breaks in the tree and adapting routing strategy to the changes in the workload.

A later version of Elvin system [15] also deals with mobile clients that periodically disconnects by introducing the notion of a proxy that is responsible with maintaining the persistency of the events and delivering them to the clients upon reconnection.

The Toronto Publish-Subscribe system (ToPSS) [11] focuses on developing an efficient content-based publish-subscribe system for high speed event notification. They also introduce the notion of approximate matching-based, which tries to take into consideration the uncertainty in the subscriber policies (like location of the client).

A somehow different problem is attacked by Huang and Garcia-Molina in [9]. They consider the case of wireless ad-hoc networks with a dynamic topology and strive to provide a reconfigurable routing scheme that will find the best configuration of routing the notifications to the mobile clients. A similar approach is used in [4], which is also an excellent review of existing publish-subscribe systems for mobile environments.

We believe that our work is orthogonal with these efforts and can be used in conjunction with them. Our main contribution is that we provide a different mechanism to define the subscribers' policies and make them usable in the presence of the dynamic resources of the mobile clients, while maintaining a high generality of the system and good performance and scalability.

Ubiquitous computing is a fast growing area for research. Several approaches have been proposed to provide intelligent environments based on embedded devices. The EasyLiving project [3] aims at providing an intelligent environment by integrating the embedded and heterogeneous devices into a coherent user experience. They propose a new middleware called InConcert which provides asynchronous message passing, machine independent addressing and XML based messages protocols. The Event Heap approach is proposed in [10] in order to accommodate interactive workspaces (systems composed of a large number of embedded devices). In that project, an extended tuple-space Linda like model is proposed. The model can be made similar to publish-subscribe using pushing mechanisms at the server side.

While there is not a consensus among the researchers on the communication model that has to be used for ubiquitous environments, the majority of the proposed methods share a lot of similarities with publish-subscribe model. The work reported in this paper can be used as a starting point in investigating how we can use more complex policies for stateful subscribers in intelligent environments.

4. STATEFUL PUBLISH-SUBSCRIBE SYSTEMS

The architecture for the SUBLIM system is based on a distributed network of servers in order to support a large number of subscribers. We assume that the subscribers are using wireless devices (such as PDAs). The servers are running on the access points of the wireless network, so that each server will be responsible for the subscribers connected to that access point. In Figure 1 we present the global architecture for our system, where WS stands for Wireless Subscriber. The solid lines represent wired network connection, while the dashed ones represent wireless connections. As we can notice, SUBLIM servers are running on the base stations, with wireless devices as subscribers. Publishers are also connecting to a SUBLIM Publish server that has as subscribers the servers running on the base stations. In the current implementation, the Publish Server is a simple router that forwards all of the messages to the SUBLIM Servers running on the base stations.

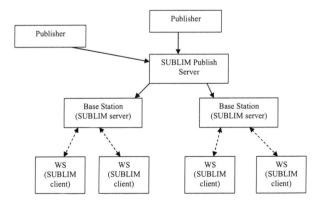

Figure 2: The SUBLIM overall architecture

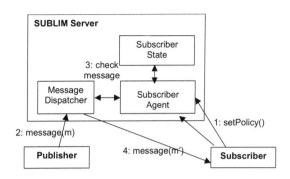

Figure 1: The architecture of the SUBLIM Server.

We will now present the abstract model used for building stateful publish-subscribe systems. The subscribers compose a community, or more specifically a S-community, where S is the SUBLIM server. To differentiate between specific subscribers, the server maintains some state for each member of the community. More formally, a S-Community is a tuple {S,C, ST, P, M}, where S is the server of the community, C is the set of subscribers, ST is a mutable set {ST_x | x ∈ C} of subscribers' state, (one per each subscriber in C), P is the set of policies, (one for each subscriber in C) and M is the set of messages that can be received by the subscribers. The state is a tuple-space and can be composed of any number of variables, describing the subscriber's properties. Each variable is uniquely identified by its name. The variables can be automatic, in the sense that their values are maintained by the SUBLIM system, or manual, maintained by the policy itself.

The SUBLIM client is a component that runs on each subscriber. The SUBLIM client is responsible for maintaining the automatic variables (as detailed in Section 6.2) and dealing with the mobility (section 4.4).

When a message arrives, the arrived() event is triggered and the policy specifies the operations that have to be performed. In our current implementation the operations are specified in a C++-like syntax. In addition to all the operations supported by C++, the policy can invoke a number of system operations that are implemented by the server. The currently supported grammar is presented in Figure 3. There are three operations associated with the handling of the properties maintained in the subscriber state (getProperty, setProperty and removeProperty). The automatic properties are read only, so the setProperty and removeProperty operations will do nothing. The policy can also specify how to create a message, using the createMessage operation, and how to manipulate the content of a message. A detailed example of such a policy is given in Section 5.

4.1 Server Architecture

The architecture of the server is presented in Figure 2. The subscriber creates the policy and sends it to the server using a setPolicy() command (Step 1 in Figure 2). The server generates an agent, which will be associated with this subscriber. Each time a message arrives at the server from a publisher, an arrived event is triggered at every connected participant.

The main difference in the design of the SUBLIM server compared with other publish-subscribe servers is the way the policies are written and interpreted by SUBLIM. The policy is translated into C++ and then linked together with the needed SUBLIM libraries in a dynamic link library (DLL). The server attaches the DLL and then uses the function provided by the DLL in order to make the decision. Thus, the server is very flexible and able to handle any number of heterogeneous policies in an efficient manner.

When a message m arrives (Step 2), it is processed by the Message Dispatcher, then the Subscriber Agent checks the message (Step 3) and decides if the message m′ (which can be m or a modified version of m) should be sent to the subscriber or not, according to its policy (Step 4).

4.2 Deploying of SUBLIM

From technical implementation point of view, the servers can run on any machines accessible from wireless devices. However, we assume that the SUBLIM servers will be deployed on the base stations, so that each server will have to support a low number of subscribers, limited to the number of devices connected to one base station. This is a natural solution for wireless environments, but can constitute a limitation for mixed wired-wireless environments, where both wired and wireless subscribers need to be supported. We designed our system in order to facilitate an incremental deployment. For wired subscribers, or for subscribers that do not need a stateful policy, any of the current systems (such as Gryphon [1], or Sienna [8]) can be used. A state, and therefore the support for a stateful policy, will be maintained only for subscribers that need such advanced features (like in the example presented in Section 5), or based on special subscriptions.

4.3 Security and Fairness

Since the policies are actually executed on the server site on behalf of the clients, the resources of the server (in terms of memory capacity, for example) can be consumed in an unfair way by malicious or bogus policies. When the policy is translated in C++ all of the possible exceptions are caught. No system calls or pointers are allowed, so there are no possible side effects. Each client is assigned one thread, so all of the clients will receive a fair treatment. In Section 6 we will give detailed experimental results on how well the SUBLIM server is supporting bogus or malicious clients. We also implemented an asynchronous mechanism that allows the server to automatically terminate the evaluation of a policy and return a NULL message if a certain timeout expires and the policy was not evaluated. The timeout is configurable and is set in the current implementation to 500 ms.

4.4 Supporting Mobility

We cannot assume that one client will stay connected to one SUBLIM server for the duration of the application. The SUBLIM client monitors all the time the available base stations. When it detects that the current base station is no longer available, it will select a new SUBLIM server based on the newly available base stations. In order to maintain the current state, the SUBLIM system implements a hand-shake mechanism. Each client remembers the address of the last server it was connected to. When the client reconnects to a new SUBLIM server A, it will send in the authentication message the address of the last server B it was connected to. The server A will contact B and will retrieve the state of that particular client. Since the two servers are assumed to have a wired connection, the state is retrieved extremely fast (less than 5 ms in a LAN environment, 100 ms in a WAN), so it will be completely transparent to the client.

5. CASE STUDY: INTENSIVE CARE UNIT IN A LARGE HOSPITAL

Consider for example a large hospital with an Intensive Care Unit. The patients are monitored continuously using specialized devices. For simplicity, we assume that the monitoring devices generate, for each patient, the following types of data: patient ID, pulse (P) and electrocardiogram (E).

Each doctor is responsible for a certain number of patients, identified by the patient IDs. The doctor wants to receive the messages according to the following policy, denoted by $P_{ICU\text{-}Stateless}$:

Operations on the associated state	
getProperty(name)	Returns the value of the specified property, or null if the property is not currently defined in the state
setProperty(name, value)	Adds a property with the specified name and value
removeProperty(name)	Remove the property with the specified name from the state
Operations on messages	
getData(m)	Get the data associated with a message
setData(m,p)	Set the data for this message
createMessage()	Create an empty message

Figure 3: The system operations

1. The doctor is interested in receiving all of the data (P and E) for the patients he is responsible for.

2. If the Pulse of a patient goes beyond a certain threshold, he is interested in getting all of the information for that patient, even if the patient is not his responsibility.

This policy will work very well in a wired environment, where the issues of low quality network connection, display size or energy supply do not exist. It is the thesis of this paper that, in a wireless environment, the current publish-subscribe paradigm will not be able to handle the dynamic changes in the quality of connection or energy supply. Intuitively, if many packets are sent to a client with a low bandwidth network connection, it might happen that, due to the congestion of the network, the client will not receive the most important information, (the pulse in this case) because the packets are dropped or delayed long enough to make the information irrelevant. The same argument can be made for the battery utilization. Receiving large amounts of data when the battery is low will reduce dramatically the availability of the service in real-life conditions.

Ideally, the doctor would like to receive all of the data for the patients he is responsible for, if he has a good network connection and the battery is charged. However, if the network connection deteriorates, he would like to be sure that he receives the most important data (such as the pulse) and maybe compressed versions of the least important data, even with loss of information such as the E. The same thing applies for the battery utilization. Similarly, the doctor will want to receive as much information as possible when the battery is charged, but only the most important data when the battery is low. Moreover, if the battery is very low, he would like only a summary of the state of the patients, at a regular interval of time. This leads to the following formulation (in an informal manner) of the policy P_{ICU}, suitable for mobile systems:

R1: The doctor is interested in receiving all data (P and E) for the patients he is responsible for if the network connection is good and the battery is charged. The electrocardiogram should not be bigger than the display size of the device used by the doctor.

R2: If the battery utilization goes beyond 75%, the doctor is interested in obtaining P and compressed versions of the electrocardiogram (E), with the compression ratio dependent on battery utilization. When the battery goes beyond 25%, the doctor is only interested in the pulse (P) every five minutes.

R3: If the network connection deteriorates, the doctor is interested in getting the pulse P and compressed versions for the electrocardiogram (E).

R4: If the pulse of the patient goes beyond a certain threshold (let us say 50), the doctor is interested in getting information about that patient according to rules **R1-R3**.

The state for this policy has three automatic variables: Battery, NetworkConnection and DisplaySize. We will detail in Section 6 how the system maintains the values for these variables; for now we can assume that the values are accurate.

An implementation for P_{ICU} is presented in Figure 4. Each of the rules is followed by comments (in italic) that, together with the following discussion, should provide enough understanding of the nature of our policies.

```
arrived(Message m, State s) :-
/* First get the values for the automatic variables */
1.      battery=s.getProperty("Battery");
2. networkConnection=s.getProperty("NetworkConnection");
3.      displaySize=s.getProperty("DisplaySize");
/* Create a new message and obtain the data associated
with the received message */
4.      Message newMessage=createMessage();
5.      pulse=m.getPulse;
6.      patientID=m.getPatientID;
7.      if (pulse > 50 and !isResponsable(patientID))
/* The doctor is not responsible for this patient and the
pulse is bigger than 50. Just ignore the message*/
8.          return NULL;
9.      endif
/* If the battery is very low, send only the pulse every five
minutes */
10.     if (battery<0.25)
11.         currentTime=getTime();
12.         oldTime=s.getProperty("Timestamp");
13.         if (oldTime==NULL)
/* The variable does not exist. Initialize it now */
14.             s.setProperty("Timestamp",currentTime);
15.             return NULL;
16.         endif
17.         if (currentTime-oldTime>5*60)
18.             s.setProperty("Timestamp",currentTime);
19.                 newMessage.setPulse(pulse);
20.                 return newMessage;
21.         else
22.                 return NULL;
23.         endif
24.     endif
25.     newMessage=
        compress(m,battery,displaySize,networkConnection);
26.     return newMessage;
```

Figure 4: The implementation of P_{ICU}.

The first three lines of the policy deal with getting the current values for the automatic variables. We then create a new message and obtain the data associated with the received messages (Rules 4-5). If the doctor is not responsible for this patient and the pulse is good, the message is just discarded by returning a NULL message (again, details on how the policy determines if the patient is within the responsibility of the doctor are given in Section 6), as shown in Rules 8-9.

If the battery is very low (below 25%), only a part of the message (the pulse) will be send to the subscriber every five minutes. This is done by using a manual variable, Timestamp, as shown in Rules 12-24.

Otherwise, the content of the message is changed based on the current values of the battery, network connection and display size. In the current implementation, the SUBLIM server provides the **compress** method, but customized methods can also be implemented in the subscribers' policies.

6. Implementation and performance

In this section we present some details of the current implementation in order to provide the reader with a better understanding of our architecture. We will focus on two main points: i) how the server creates the agent that will interpret the subscriber policy and ii) how the server maintains the variables in each subscriber's state, particularly the network connection.

6.1 Interpreting the Subscriber Policy

Upon connection to a SUBLIM server, the subscriber needs to specify its policy. The server translates the policy into C++ code, compiles it into a dynamic link library (DLL) and then loads the DLL into memory. The DLL exports only one function, which will be invoked by the arrived event. The input for this function is the message, encapsulated in a special class (Message) and the current state of the subscriber. The output of this method will be the message that it is sent to the subscriber. A NULL message means the subscriber is not interested in this message. Compiling the policy into an executable code (DLL) only once, and loading the DLL into memory significantly improves the performance of the whole system

6.2 Maintaining the Automatic Variables in Subscriber State

The automatic variables from the state of each client are maintained by the SUBLIM server using a simple protocol of communication with the subscriber. Periodically (30 seconds in the current implementation), the client sends to the server the new values for the variables, in a state package. Variables such as battery utilization, display size, etc. are obtained directly at the subscriber site and sent back to the SUBLIM server. However, for the network connection variable, additional computation needs to be done at the server site. The server records the number of messages sent to subscriber S during an interval of time, nMessagesSent. Subscriber S sends back to the server, in the state package, the number of messages it received in that interval, nMessagesReceived. The network connection variable is simply the ratio nMessagesReceived/nMessagesSent. Intuitively, a high value (close to 1) of this ratio means that the bandwidth can support the current traffic to that subscriber. If the ratio decreases, it means that the messages are lost or delayed in the network, so the subscriber policy needs to apply special measures, like the ones described in Section 5.

We cannot assume that one client will stay connected to one SUBLIM server for the duration of the application. The SUBLIM client monitors the available base stations permanently. When it detects that the current base station is no longer available, it will select a new SUBLIM server based on the newly available base stations. In order to maintain the current state, the SUBLIM system implements a hand-shake mechanism. Each client remembers the address of the last server it was connected to. When the client reconnects to a new SUBLIM server A, it will send in the authentication message the address of the last server B it was connected to. The server A will contact B and will retrieve the state of that particular client. Since the two servers are assumed to have a wired connection, the state is retrieved extremely fast (less than 5 ms in a LAN environment, 100 ms in a WAN), so it will be completely transparent to the client.

6.3 Performance

The stateful model for publish-subscribe systems is a very generic and flexible model, allowing for a large variety of applications and policies. The performance of the system cannot be evaluated for a generic case, since it is affected by the specific application, the format of the data that has to be sent to the subscriber, the state of the subscriber and the policy of the subscriber.

Our current server fully implements the stateful model allowing for flexible deployment of various applications. For the performance evaluations, we will use the Intensive Care Unit example, with the policy specified in Section 5. The subscribers are running on wireless devices using PDAs and laptops. For the evaluation, we used three different types of PDAs, all of them running Windows CE 2.0: an HP Jornada 540, with 32 MB of RAM and a SH3 processor at 133 MHz, a Cassio Cassiopeia E-125 with 32 MB of RAM and a MIPS processor at 150 MHz and a Compaq iPaq with a strongARM processor at 300 MHz. The results are very similar for all of the three devices, so we will only show the results obtained using the Jornada device. The SUBLIM server is running on a dual Pentium Xeon at 2.8GHz with Linux 2.4.0 kernel. The publisher is connected directly to the SUBILM server and can send messages at a variable rate. The electrocardiogram is a JPEG image, and the server is using the Independent Group JPEG library [13] in order to compress the image according to the policy P_{ICU}.

We will first study what is the overhead introduced by maintaining the state for each subscriber. The test scenario comprises n subscribers and one publisher, generating messages at a variable rate. When the server receives the first message from the publisher, it records this time as *start* time. The server records the *ending* time when it finishes processing all the messages it has received from the publisher. The duration of the experiment is determined by the difference between the end time and the start time.

We compare against a "blind" publish-subscribe system, in which there are no agents associated with the subscribers and no states maintained. Each subscriber receives all the messages sent by the publisher; the server acts just like a router in this case. As the metric we consider the duration of the experiment, as defined above and we define the overhead as: $O=(D_{Full}-D_{Blind})/D_{Blind}*100$, where D_{Full} is the duration of the experiment for the stateful system and D_{Blind} is the duration of the experiment for the "blind" publish-subscribe system. The average size of a message is 100 KB and the duration of each experiment is 5 minutes. The rate varies between 30 and 120 mpm (messages per minute). For this experiment, we run the subscribers on wired devices.

The results are shown in Figure 5 averaged over 20 executions for each case. As we can notice, the overhead is 0 even for a large number of subscribers (220), when the rate of the published messages is not very high. However, when the rate increases to 120 mpm, the overhead becomes important when the number of subscribers is high (more than 140). This is a very promising result showing that the overhead is almost negligible, even for relatively complex stateful policies like P_{ICU}, for all practical scenarios. This is because the number of subscribers connected to one SUBLIM server is limited by the number of wireless devices connected to one base station, which is usually smaller than 100 and cannot increase due to the physical limitations of the wireless channels.

We will now present how the SUBLIM server behaves under dynamic changes of the bandwidth for a wireless subscriber. In our

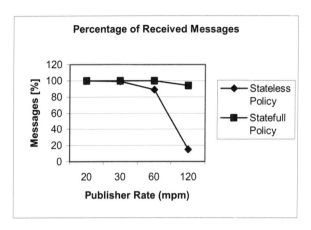

Figure 6: Percentage of received messages.

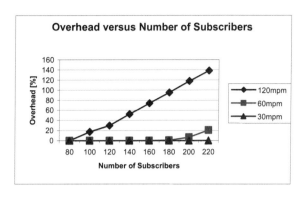

Figure 7: Overhead versus number of subscribers.

experiments, we have one subscriber, one patient and the publisher is able to send messages at a variable rate. We are measuring how many messages are received by the subscriber compared with the number of messages sent by the publisher, with stateful policies and stateless policies (P_{ICU} and $P_{ICU\text{-Stateless}}$ from Section 1). The bandwidth varies randomly between 300 kbs and 3Mbs. The average size of a message is 100 KB and the duration is 5 minutes for each experiment. As in the previous experiment, the publishing rate varies between 20 and 120 mpm (messages per minute). In this experiment, the battery of the subscriber remained charged the whole time, so the number of messages that match the subscriber policy is the same in both cases. The results, averaged over 20 executions, are presented in Figure 6.

As we can notice, when the rate of the publishing messages increases, the percentage of received messages by the subscriber decreases very fast for the stateless policy. This is because the system has no way to adapt itself to the dynamic changes in the bandwidth, a well-known attribute for wireless environments. For the stateful case, the percentage is very high, even for high rates for publishing messages.

Our next measurements are related to the effect of the stateful policy on battery duration for a device. We present in Figure 7 the time needed for the battery to reach the 0 power level (the battery life), under stateful and stateless conditions. No changes in the bandwidth were performed during this experiment. The first line in Figure 7 is the battery life when no messages are published. The conditions are

the same as in the previous experiment: one subscriber, one publisher, one patient.

As we notice, we can extend the battery life by as much as 31%, dramatically increasing the usability of publish-subscribe systems in wireless environments.

We finally present in Figure 8 the results for the robustness of our system in the presence of bogus/malicious agents. We use the same setup as in the first experiment, namely one SUBLIM server running on a dual Pentium Xeon at 2.8GHz with Linux 2.4.0 kernel and wired subscribers. As we specified in Section 4.3, the language used for writing the policies does not allow for use of pointers or any kind of loops, so the possibility of attacks against the server due to malicious/bogus agents is greatly reduced. However, we reduced these limitations in order to get experimental results for the robustness of the system even by allowing some subscribers to specify a policy that will contain infinite loops. A **malicious** agent is defined as an agent who sets a policy that contains an infinite loop. We were interested to see how the overhead of the system is affected by malicious agents. We basically run the same experiments as the ones from Figure 5 but we introduced a variable number of malicious agents. The results are presented in Figure 8a. We varied the percentage of malicious agents from 0 to 10. The results for the same experiments, but with a timeout value of 500 ms (as also explained in Section 4.3) are presented in Figure 8b.

As we can notice, having an asynchronous timeout mechanism is a very effective method to achieve a high robustness of the system under heavy load (200 subscribers and 240 mpm) and with a high number of malicious agents.

7. Conclusions

There are two major contributions of this paper. First, we introduce the notion of stateful publish-subscribe systems and argue that the current publish-subscribe systems are not suitable for mobile systems. A novel approach for expressing the conditions and building the agents using a run-time compiling technique is presented. This technique allows for the conditions to interact with the current state of each subscriber and decide whether or not to send it to the subscriber and how to modify the message if necessary.

Second, we present a motivational example where our techniques can be beneficially used. Moreover, our approach can be incorporated in the current publish-subscribe systems in an

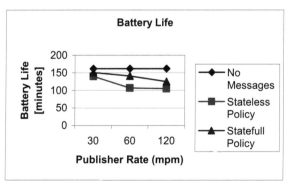

Figure 5: Battery life

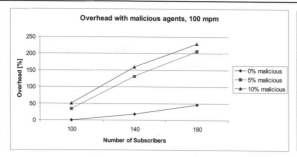
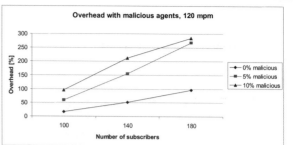

Figure 8a: Overhead with malicious agents, no timeout

	120 mpm			240 mpm		
Number of subscribers	0% malicious	5% malicious	10% malicious	0% malicious	5% malicious	10% malicious
100	17.3	14.6	13.6	119.66	118.8	123.33
140	52.3	51	50.6	206.6	204.44	203.33
180	95.3	93	94.6	288.66	290.55	287.77
200	118	116	117	334.33	328	333.88

Figure 8b: Overhead with malicious agents, 500 ms timeout

incremental manner, by allowing for some of the subscribers to maintain a state at the server side. Experimental results demonstrate the good performance, scalability and robustness of our approach.

More information about this work is provided at the web site *http://www.cs.rutgers.edu/~ionescu/Sublim.*

8. REFERENCES

[1] M. K. Aguilera, R. E. Storm, D. C. Sturman, M. Astley, and T. Chandra. Matching events in a content-based subscription system. In 18th ACM Symposium on Principles of Distributing Computing (PODC), 1999.

[2] A. Campailla, S. Chaki, E. Clarke, S. Jha and H. Veith. Efficient filtering in publish-subscribe systems using binary decision diagrams. In International Conference on Software Engineering, 2001.

[3] B.L. Brumitt, B. Meyers, J. Krumm, A. Kern, and S. Shafer. EasyLiving: Technologies for Intelligent Environments". In Handheld and Ubiquitous Computing, 2nd Intl. Symposium, September 2000.

[4] Mauro Caporuscio, Antonio Carzaniga and A. L. Wolf. Design and Evaluation of a Support Service for Mobile, Wireless Publish/Subscribe Applications. In IEEE Transactions on Software Engineering, vol. 29, December 2003.

[5] G. Cugola, E. Di Nitto and A. Fuggetta. The JEDI event-based infrastructure and its application to the development of the OPSS WFMS. In IEEE Transactions on Software Engineering, vol. 27, pp. 827-850, Sept. 2001.

[6] G. Cugola, G.Picco and A. Murphy. Towards dynamic reconfiguration of distributed dynamic reconfiguration of distributed publish-subscribe middleware. In 3rd International Workshop on Software Engineering and Middleware (SEM 2003), May 2002.

[7] G.Cugola and E. Di Nitto. Using a publish-subscribe middleware to support mobile computing. In Proceedings of the Workshop on Middleware for Mobile Computing, November 2001.

[8] A. Carzaniga, D. S. Rosemblum and A. L. Wolf. Achieving scalability and expressiveness in an internet-scale event notification service. In 19th ACM Symposium on Principles of Distributing Computin PODC, 2000.

[9] Yongqiang Huang and Hector Garcia-Molina. Publish/Subscribe Tree Construction in Wireless Ad-Hoc Networks. In Proceedings of the 4th International Conference on Mobile Data Management (MDM), 2003.

[10] Brad Johanson and Armando Fox: The Event Heap: A Coordination Infrastructure for Interactive Workspaces. In Fourth IEEE Workshop on Mobile Computing Systems and Applications, June 2002.

[11] H. Leung and H.-A. Jacobsen. Subject-Spaces: A state persistent programming model for publish-subscribe systems. Technical Report, University of Toronto, September 2002.

[12] Keryx homepage, http://keryxsoft.hpl.hp.com

[13] Independent JPEG group home page, http://www.ijg.org/

[14] B. Segall and D. Arnold. Elvin has left the building: A publish-subscribe notification service with quenching. In Proceedings of the Australian UNIX and Open Systems User Group Conference, 1997.

[15] P. Sutton, R. Arkins and B.Segall. Supporting Disconnectedness – Transparent Information Deli very for Mobile and Invisible Computing. In Proceedings of the IEEE International Symposium on Clustering Computing and the Grid, May 2001.

Reputation-based Wi-Fi Deployment Protocols and Security Analysis*

Naouel Ben Salem   Jean-Pierre Hubaux
Laboratory of Computer Communications and Applications
Swiss Federal Institute of Technology Lausanne (EPFL)
Switzerland

{naouel.bensalem,jean-pierre.hubaux}@epfl.ch

Markus Jakobsson[†]
School of Informatics
Indiana University at Bloomington
Bloomington, IN 47406, USA

mjakobsson@rsasecurity.com

ABSTRACT

In recent years, wireless Internet service providers (WISPs) have established thousands of WiFi hot spots in cafes, hotels and airports in order to offer to travelling Internet users access to email, web or other Internet service. However, two major problems still slow down the deployment of this kind of networks: the lack of a seamless roaming scheme and the variable quality of service experienced by the users. This paper provides a response to these two problems: We present a solution that, on the one hand, allows a mobile node to connect to a foreign WISP in a secure way while preserving its anonymity and, on the other hand, encourages the WISPs to provide the users with good QoS. We analyse the robustness of our solution against various attacks and we prove by means of simulations that our reputation model indeed encourages the WISPs to behave correctly.

Categories and Subject Descriptors: C.2.1: Wireless communication, Network communications.

General Terms: Design, Security, Theory.

Keywords: WiFi networks, Reputation systems, Roaming, billing, Security, QoS, Protocols.

1. INTRODUCTION

Wireless data services based on cellular networks, such as GSM/GPRS, provide users with very good coverage. However, they have several intrinsic and well-known drawbacks: the offered bitrates are relatively low (and this is unlikely to change with the Third Generation), and the deployment of new features is hampered by several factors such as the large size and oligopolistic behavior of the operators, their willingness to provide homogeneous service, and the huge upfront investment; in addition, very often, a user located in his home country is not allowed to obtain service from the competitors of his home network.

The deployment of wireless networks such as WiFi in unlicensed frequencies makes it possible to envision a substantial *paradigm shift*, with very significant benefits: much higher bandwidth network, deployment based possibly on local initiative, higher competition, and much shorter time-to-market for new features. This may, in turn, pave the way for new types of services, whether these require higher bandwidth, lower per-bit costs, reduced energy consumption for the mobile nodes or higher reliance on fast-changing and locally provided content.

The current, rapid deployment of hot spots reveals the strong potential of this approach. However, two major problems still need to be solved. The first problem is the provision of a seamless roaming[1] scheme that would encourage small operators to enter into the market. This is a fundamental issue for the future of mobile communications. Indeed, without an appropriate scheme, only large stakeholders would be able to operate their network in a profitable way, and would impose a market organization very similar to the one observed today for cellular networks; one of the greatest opportunities to fuel innovation in wireless communications would be missed. The second problem is the guarantee of a good quality of service provision to the users.

This paper provides a response to these two challenges. By appropriately unbundling the major functions of the network, it institutes a virtuous cycle of deployment and usage: Wireless Internet Service Providers (WISP) will be encouraged to deploy their network and will be confident that mobile users registered with other WISPs will pay for the service they receive; likewise, users will be assured that the WISPs are under the scrutiny of all the other users (including the roaming ones), and that they will be informed about their degree of satisfaction.

As we will see, the solution is relatively simple, provided that the roles of the different entities are clearly defined. We describe these entities in detail, along with the security protocols and the charging mechanism. In order to facilitate user acceptance, the proposed solution minimizes user involvement: once the mobile device has been initialized, it can make all decisions autonomously.

[†]Work done while at RSA Laboratories.

*The work presented in this paper was supported (in part) by the National Competence Center in Research on Mobile Information and Communication Systems (NCCR-MICS), a center supported by the Swiss National Science Foundation under grant number 5005 − 67322

---

[1]Note that by roaming we designate the operation of obtaining service from different operators, and not the handoff between access points (managed by the same provider or by two different providers). The handoff problem is out of the scope of this paper.

One of the major goals of this work is to build up trust between mobile users and WISPs. For this reason, we provide a detailed threat analysis and we show that the proposed protocols can thwart rational attacks and detect malicious attacks (we define these terms in Subsection 2.2).

The rest of the paper is organized in the following way: In Section 2 we present the system and trust models and we give an overview of the proposed solution. In Section 3, we describe the details of the protocols. We study the security of the protocols and analyse some interesting aspects of the solution in Section 4. In Section 5, the simulations are described and the results are analyzed. Finally, we present the state of the art in Section 6 and we conclude in Section 7.

2. SYSTEM MODEL

In this paper, we consider a mobile node MN that wants to connect to the Internet via a neighboring hot spot (i.e., a hot spot that is within its power range); we assume the hot spot to be managed by a Wireless Internet Service Provider (WISP) that we denote by S (see Figure 1). MN is affiliated with its home WISP H[2] with whom it has an account and shares a symmetric key k_{HM}. We assume that all the messages exchanged between MN and H go through S. Note that it is possible to have $S = H$.

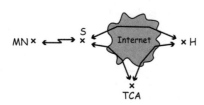

Figure 1: The system model.

All WISPs in our model are registered with the trusted central authority TCA that creates for each of them a public/private key pair and a certificate of their public key and of their identity. In a "grassroots" vision, the TCA would be a federation of WISPs, who join forces to centralize a few strategic functions. In a more conventional vision, the TCA can be under the control of a world-wide organization such as a quality control company, a certification company, or a global telecommunications operator. TCA can be distributed, as certification companies are, to avoid being a bottleneck.

In this paper, we present a reputation based mechanism that, on the one hand, allows MN to evaluate the behavior of the WISPs and, on the other hand, encourages the WISPs to provide the users with good QoS. Each WISP in our model has what we call a *reputation record* that represents an evaluation of its behavior and that is generated and signed by TCA. The choice of the initial reputation record of a WISP is discussed in Section 5.

In order to make sure that the mobile nodes pay for the service they receive, we also propose a credit-based micropayment scheme (see Subsection 3.1.1) that is highly inspired from the PayWord scheme [20]. Our solution takes into account the fact that MN is a ressource restrained mobile device and therefore has much less computing and storage resources than TCA, H or S.

---

[2]The solution works even if H does not operate hotspots itself.

2.1 Assumptions

We make the following assumptions in this paper:

- The public key of TCA is known by all other entities.

- H and S can use their public keys to establish a temporary symmetric key k_{HS}. We assume that this key is generated prior to the execution of our set of protocols.

- S is able to predict the QoS it can offer MN. We will discuss this issue more in detail in Subsection 5.3.

- The backbone is a commodity; the rewarding of the backbone operator should follow already established practices and techniques, and will not be addressed in this paper (we assume that S, H and TCA have an appropriate agreement to have connectivity - e.g., a flat rate subscription).

- In this paper, we do not consider the handoff problem: A mobile node that moves out of the range of a hot spot stops using that hot spot as access point and initiates a new connection with one of the new neighboring hot spots.

2.2 Trust and adversarial model

We consider an attacker $\mathcal{A}$ that wants to perform an attack against our protocols (see Subsection 4.1 for the list of attacks). $\mathcal{A}$ can be a mobile node or a WISP. We assume that:

- TCA never cheats and is trusted by the other parties for all the actions it performs.

- The WISPs (here S and H) are rational and therefore they cheat (i.e., perform one of the attacks presented in Subsection 4.1) only if it is to their advantage (i.e., they gain something - in terms of money - from cheating). This assumption is reasonable because a WISP is likely to be motivated by economic incentives, and would not be inclined to disrupt the communication of mobile nodes (who could simply choose another WISP if this were to occur).

- MN may be malicious and therefore it can cheat (i.e., perform one of the attacks presented in Subsection 4.1) even if there is no gain from cheating (this implicitly assumes that MN can also perform rational attacks).

- MN trusts H for managing its account.

- Several attackers can collude and share information (possibly their secret keys) to perform more sophisticated attacks.

Confidentiality of data is not an issue in our case, so we do not consider passive attacks where the attacker eavesdrops the data exchanges between two parties. Note that this is an orthogonal issue that is easily addressed using standard security techniques.

We consider exclusively attacks performed against the different phases of our protocols, meaning that we do not consider other arbitrary attacks like DoS attacks based on jamming for example.

In this paper, we want to study the effect of rational and malicious attacks on our set of protocols. Our goal is to make sure that our solution thwarts rational attacks, detects malicious attacks and, if possible, identifies the attacker.

2.3 Rationale of the solution

When *MN* wants to connect to the Internet, it identifies the neighboring WISPs[3] and contacts them (see Figure 2). Each WISP sends to *MN* an offer that contains its reputation record, the QoS it proposes and the price it asks for. Then, *MN* selects the WISP *S* that proposes the best offer and verifies its identity. *S* also verifies, with the help of *H*, that *MN* is a valid node. *MN* and *S* establish a contract, inform *TCA* and *H* about it and establish a secure session by setting up a symmetric key k_{MS}.

This secure session is divided into parts. During the *i*-th part, *MN* sends a payment proof for the *i*-th part of the service and *S* provides that part of the service. The payment proofs and the services are secured using the shared key k_{MS}.

At the end of the connection, *MN* assesses the QoS it received, compares it to the QoS advertised by *S* during the session setup and informs *TCA* about its *satisfaction level*. *S* also sends the payment proof(s) to *H* which charges *MN* (by manipulating its account) and remunerates *S* according to the received information.

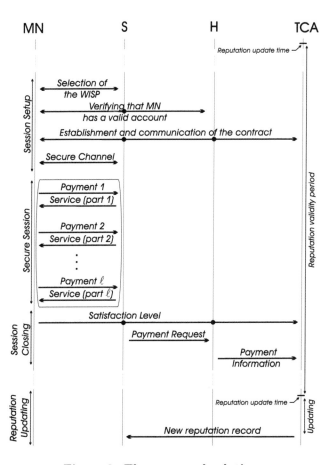

Figure 2: The proposed solution

TCA collects the feedback about the different WISPs, updates periodically the reputation records according to the collected information and provides the WISPs with their new reputation records.

---

[3]Note that we refer to the access points using the identities of the WISPs that are managing them.

3. PROPOSED SOLUTION

3.1 Basic mechanisms

3.1.1 Micro-payment scheme

As already mentioned in Section 2, the payment scheme we use in this paper is highly inspired from the PayWord scheme [20]: During the session setup, *MN* generates a long fresh chain of paywords $w_0, w_1, ..., w_n$ by choosing w_n at random and by computing $w_i = h(w_{i+1})$ for $i = n-1, n-2, ..., 0$, where h is a one-way hash function and n is the maximum number of payments that *MN* can send to *S* during the session. Then, *MN* reveals the root w_0 of the payword chain (which is not considered as a payword itself) to *S*, *H* and *TCA*.

During the session, *MN* sends (w_i, i) to *S* as a payment proof for the *i*-th part of the service. *S* can easily verify w_i using w_{i-1} that is known from the previous micropayment or from w_0 if $i = 1$. At the end of the service provision, *S* contacts *H* and presents the last payment (w_ℓ, ℓ) it received. *H* verifies the validity of w_ℓ, pays *S* the amount corresponding to ℓ paywords and charges *MN* for that amount by manipulating its billing account.

We use this micropayment scheme because it allows an offline verification of the payment proofs and because of its low computational and storage cost for the mobile nodes.

3.1.2 Authentication of MN by H

As stated in Section 2, all communication between *MN* and *H* goes through *S*. Therefore, in order to preserve the anonymity of *MN* regarding *S*, we use the following authentication mechanism, that is commonly used in the industry (e.g., SecurID [11]): When *MN* gets affiliated with *H*, the two parties share a random seed *s* that represents the input to a pseudorandom generator. The output is a random number *tag* that is 30 to 50 bits long. *H* keeps a small window (e.g., 50 entries) of upcoming tags for each mobile node and maintains the pairs (*tag; node's identity*) in a sorted database. Upon receipt of a given *tag*, *H* searches its database, retrieves the pair (*tag;identity*) and identifies *MN*. In case of collision (i.e., more than one pair contains the random number *tag*), *H* asks *MN* to send the next tag value.

3.2 Details of the protocols

3.2.1 Selection of the WISP

When it wants to obtain Internet access, *MN* scans the spectrum, identifies the neighboring WISPs and asks them an offer by broadcasting the following request message:

$$OfferReq = [ReqID, n_M] \qquad (1)$$

where *ReqID* is the request identifier and n_M is a nonce generated by MN. Each WISP *W* willing (and able) to provide service at that time responds by a signed offer *Offer_W*:

$$W \rightarrow MN : Offer_W, S_{pk_W}(Offer_W, n_M) \quad \text{where}$$
$$Offer_W = [W, RR_W, AQ_W, P_W, Cert(W)] \qquad (2)$$

where RR_W is the most recent *reputation record* of *W* (signed by *TCA*), AQ_W is the QoS it advertises[4], P_W is

---

[4]*W* may advertise a QoS that is higher than the real QoS (RQ_W) it is able to offer to *MN*. The consequences of such a behavior are studied in Section 5.

the price it is requesting for each part of the service (see Subsection 3.2.4), pk_W is its private key and $Cert(W)$ is the certificate of its public key PK_W.

Upon receipt of the offers, MN verifies the freshness of n_W and identifies the best offer. This choice depends on the relative importance that MN gives to the parameters R_W, Q_W and P_W (as shown in Section 5, these parameters can depend on the application MN intends to run) and should be made by a software agent to automate the process and avoid human involvement. More sophisticated schemes (e.g., auctioning) can also be envisioned.

Then, MN verifies the certificate and the signature of the WISP that proposed the best offer. If the verification is incorrect, MN checks the second best offer and so on. We denote the selected WISP by S.

3.2.2 Verifying that MN has a valid account

Before starting the session, S has to make sure that MN is a valid mobile node that is registered with a valid home WISP. As we want to preserve the anonymity of MN, the verification of MN's identity involves H and uses the authentication mechanism described in Subsection 3.1.2. We have thus the following messages exchanged:

$$MN \rightarrow S \quad : \quad \mathcal{M} = [H, tag, n_M$$
$$E_{k_{HM}}(MN, S, tag, n_M)] \qquad (3)$$
$$S \rightarrow H \quad : \quad S, n_S, \mathcal{M}, MAC_{k_{HS}}(S, \mathcal{M}) \qquad (4)$$
$$H \rightarrow S \quad : \quad TID, E_{k_{HM}}(TID, n_M, k_{MS}),$$
$$E_{k_{HS}}(TID, n_S, k_{MS}) \qquad (5)$$
$$S \rightarrow MN \quad : \quad TID, E_{k_{HM}}(TID, n_M, k_{MS}) \qquad (6)$$

(3) MN sends to S a message $\mathcal{M}$ containing, in clear, the identity of H, its current *tag* and a freshly generated nonce n_M. $\mathcal{M}$ also contains, encrypted using the symmetric key k_{HM}, the identities of MN and S, the tag *tag* and the nonce n_M.

(4) S sends to H its identity, a freshly generated nonce n_S, the message $\mathcal{M}$ and a MAC computed on both items using the key k_{HS}.

(5) H searches its sorted database, identifies MN using the *tag* sent in clear (as explained in Subsection 3.1.2), looks up the symmetric key it shares with MN and uses it to decrypt the rest of the message. Then, H re-checks the identity of MN (the identity corresponding to the tag should also correspond to the identity MN encrypted in the message) and verifies that the WISP with which MN intends to interact is indeed the WISP that sent the message.

If the message is not correct, H informs S that MN is not affiliated with it by sending a negative acknowledgement. If, on the contrary, the message verifies correctly, H generates a symmetric key k_{MS} that MN and S will use later as a session key (i.e., all the messages exchanged between MN and S during the session are secured using k_{MS}). Then, H constructs a message containing:

- in clear, a fresh temporary identifier TID for MN (TID will be used during service provision),

- TID, n_M, and k_{MS} encrypted using the symmetric key k_{HM}, and

- TID, n_S, and k_{MS} encrypted using the symmetric key k_{HS},

and sends this message to S. H maintains a table containing the correspondence between the temporary identifiers and the identities of the nodes; given TID, H can positively identify the correspondent MN.

(6) S decrypts $E_{k_{HS}}(TID, n_M, k_{MS})$, verifies that the temporary identifier in the decrypted part corresponds to the one sent in clear and compares the nonce in the decrypted part with the one generated by MN. If these verifications are correct, S removes $E_{k_{HS}}(TID, n_M, k_{MS})$ from the message and forwards the rest to MN.

MN decrypts $E_{k_{HM}}(TID, n_H, k_{MS})$ and verifies the temporary identifier and the nonce as S did. If everything is correct, MN maintains TID in memory.

Note that if $S = H$, MN sends message (3) to H and H responds with message (6).

3.2.3 Contract establishment and communication

During this phase, MN generates a long hash chain of $n + 1$ elements, computed from a randomly chosen seed w_n as described in Subsection 3.1.1. Then MN generates a contract C as follows:

$$C \quad = \quad [CID, w_0, R_S, AQ_S, P_S]$$

where $CID = [TID, S, H]$ is the contract identifier and w_0 is the root of the hash chain.

Then MN and S inform H about the contract:

$$MN \rightarrow S \quad : \quad C, MAC_{k_{MS}}(C), MAC_{k_{HM}}(C) \qquad (7)$$
$$S \rightarrow H \quad : \quad C, MAC_{k_{HM}}(C), MAC_{k_{HS}}(C) \qquad (8)$$

(7) MN sends the contract C to S, together with two MACs computed on C using the symmetric keys k_{MS} and k_{HM}, respectively.

(8) S verifies C and $MAC_{k_{MS}}(C)$ and if they are correct, it computes a MAC on C using the symmetric key k_{HS} it shares with H. Then, S sends to H the contract C and the MACs computed with k_{HM} and k_{HS}. H verifies the MACs and, if they are correct, it stores the contract C.

MN and S also inform TCA about the contract:

$$MN \rightarrow S \quad : \quad E_{PK_{TCA}}(C, k_{MT}, pad),$$
$$MAC_{k_{MS}}(E_{PK_{TCA}}(C, k_{MT}, pad)) \qquad (9)$$
$$S \rightarrow TCA \quad : \quad C, E_{PK_{TCA}}(C, k_{MT}, pad) \qquad (10)$$
$$TCA \rightarrow S \quad : \quad S_{pk_{TCA}}(C), MAC_{k_{MT}}(C) \qquad (11)$$
$$S \rightarrow MN \quad : \quad MAC_{k_{MT}}(C) \qquad (12)$$

(9) MN generates a fresh symmetric key k_{MT} that MN will use later to encrypt data for TCA (see Subsection 3.2.6). In order to prevent the key retrieval by an attacker, MN uses the probabilistic encryption by appending to the key a pseudorandomly generated bitstring *pad* (the length on the bitstring depends on the encryption algorithm used). Then, MN encrypts C, k_{MS} and *pad* using the public key of TCA, computes a MAC on this data using the key k_{MS} it shares with S and sends the encrypted data and the MAC to S.

(10) S verifies the MAC, removes it and sends C and the encrypted data to TCA.

(11) TCA decrypts the data and compares the contract C received in the encrypted data with the contract received in clear from S. If they are identical, TCA signs the contract C using its private key pk_TCA, computes a MAC on it using the symmetric key k_{MT} that is shares with MN, and sends the signature and the MAC back to S. TCA also maintains C and k_{MT} in its local database.

(12) S verifies the signature and if correct, it forwards the MAC to MN which verifies it and stores k_{MT}.

3.2.4 Service provision and payment

The session is subdivided into parts, depending on the duration or on the amount of data exchanged between MN and S. During the i-th part:

$$MN \rightarrow S \quad : \quad TID, w_i, MAC_{k_{MS}}(TID, w_i) \quad (13)$$
$$S \rightarrow MN \quad : \quad \text{i-th part of the service,}$$
$$MAC_{k_{MS}}(\text{i-th part of the service}) \quad (14)$$

(13) MN sends to S its temporary identity TID, the i-th PayWord w_i and a MAC computed on both items using the key k_{MS}.

(14) S verifies the validity of w_i by checking that $h(w_i) = w_{i-1}$, where h is the one-way hash function used by MN to generate the chain. If it is correct, S provides MN with the i-th part of the service.

3.2.5 Sending the payment request

At the end of the session, S sends to H a payment request PR that contains, encrypted using k_{HS}, the contract identifier CID, the last hash value w_ℓ it received from MN and the number ℓ of provided service parts. PR also contains, in clear, the identity of S so that H is able to retrieve the symmetric key k_{HS}.

$$S \rightarrow H \quad : \quad PR = [S, CID, w_\ell, \ell,$$
$$MAC_{k_{HS}}(S, CID, w_\ell, \ell)] \quad (15)$$

Upon receipt of PR, H verifies the validity of w_ℓ as explained in Subsection 3.1.1, retrieves the price P_S from the contract, rewards S for the ℓ parts of the service, and charges MN. H is also remunerated (see details in Subsection 3.3).

3.2.6 Sending the satisfaction level

At the end of the session, MN generates a *satisfaction level* message Sl as follows:

$$Sl = [E_{k_{MT}}(CID, QoSEval_{S,CID}, w_\ell, \ell)] \quad (16)$$

$QoSEval_{S,CID}$ is expressed by MN and compares to what extend the QoS it obtained during the session is complaint with the QoS announced by S in the offer. k_{MT} is the key MN shares with TCA.

Then, MN reports on its *satisfaction level* to TCA:

$$MN \rightarrow S \quad : \quad TID, Sl, MAC_{k_{MS}}(TID, Sl) \quad (17)$$
$$S \rightarrow TCA \quad : \quad S, CID, w_\ell, \ell, Sl,,$$
$$S_{PK_S}(S, CID, w_\ell, \ell, Sl) \quad (18)$$

(17) MN sends to S its temporary identifier TID, Sl data and a MAC computed on both items.

(18) S verifies the MAC. If it is correct, S generates a message containing CID, w_ℓ, ℓ and Sl, signs it and sends the message and the signature to TCA.

TCA verifies the signature and retrieves the key it shares with MN (using CID). Then TCA decrypts Sl, compares the CID, w_ℓ, ℓ in the encrypted data to those received in clear from S and if they are identical, TCA considers $QoSEval$ as a valid feedback. Then TCA informs H that it correctly received the feedback:

$$TCA \rightarrow H \quad : \quad Ack, S, CID, S_{PK_{TCA}}(Ack, S, CID) \quad (19)$$

(19) H verifies the signature and retrieves the identity of MN (using CID). Then, H remunerates MN a small amount of money ε, which is meant to encourage the mobile nodes sending the reports.

3.2.7 Updating the reputation record

TCA collects the information about the satisfaction levels for a given period and then, at the *reputation update time*, TCA updates the reputation record of each WISP, signs them and informs the WISPs about their new records. The new reputation record depends on the old one and on the collected information. An example is given in Subsection 5.

TCA considers the absence of feedback as negative feedback. Indeed, TCA knows that a session has been established between MN and S and that H is the home WISP of MN (see Subsection 3.2.3). TCA is thus waiting for the report from MN about its interaction with S, and not receiving it within a "reasonable" time is considered as bad feedback.

3.3 Charging and rewarding model

In the previous Subsection (3.2), we presented the details of the solution. Some of these details are related to the charging and rewarding mechanism and we summarize them in this Subsection:

- During session setup, MN generates a chain of paywords w_0, w_1, ..., w_n.

- During the secure session with S, MN sends (w_i, i) to S as a payment proof for the i-th part of the service. S can check the validity of the payment by verifying that $w_{i-1} = h(w_i)$.

- H remunerates MN a small amount ε when it receives from TCA the confirmation that MN reported on its interaction with S.

  If, at the end of the session, MN moves away from S (and therefore cannot send the feedback via S), it is still possible for MN to report on its satisfaction level to TCA via another WISP W: W includes its identity in message (18) and signs the message using its own private key. TCA then verifies the signature and informs H in message (19) about the identity of W. Then H gives both MN and W a reward (e.g., $\varepsilon/2$ for each).

- At the end of the session, S sends to H the last payment proof (w_ℓ, ℓ) it received from MN. H verifies the validity of the payword w_ℓ, charges MN the amount $P_S * \ell$ corresponding to the ℓ parts of the service and rewards S, using a well-established e-payment technique, the amount[5] $P_S * \ell - \varepsilon$. If TCA receives no report from MN, ε is handled according to some policy (e.g. it can be distributed to charity).

- The home network H is also remunerated. This can be done e.g., if MN pays a flat monthly subscription A or if MN pays an amount a per session. For sake of simplicity, we consider the second approach in this paper.

---

[5]As already mentioned, ε is the reward MN receives if it reports on its satisfaction level to TCA.

4. SECURITY ASSESSMENT

4.1 Attacks

In this Subsection, we identify the attacks that an attacker[6] $\mathcal{A}$ may perform against our protocols (see Subsection 2.2 for the trust and adversarial model). We identify the following attacks that are specific to our solution:

- *Publicity* attack: In Message 2, S advertises a QoS that is higher than the real QoS it can offer.

- *Selective publicity* attack: S performs the Publicity attack with a specific MN.

- *Denigration* attack: MN receives a good QoS from S but pretends the contrary by sending a negative report or no report at all.

- *Flattering* attack: MN sends systematically a good feedback about S's behavior to TCA. This attack makes sense particularly if $S = H$.

- *Report dropping* attack: MN sends the report but S does not transmit it to TCA.

- *Service interruption* attack: S receives the i-th payment proof from MN but does not provide the corresponding part of the service.

- *Refusal to pay* attack: MN does not send the i-th payment to S.

- *Repudiation* attack: S or MN retracts the agreement it has with other party (e.g., S asks for higher price than agreed on when the contract C was established).

We also consider general attacks such as:

- *Packet dropping* attack: $\mathcal{A}$ drops a message it is asked to forward or discards a message it is asked to generate and send.

- *Filtering* attack: $\mathcal{A}$ modifies a packet it is asked to forward or generate.

- *Replay* attack: $\mathcal{A}$ replays a valid message that was exchanged between two legitimate parties.

We do not consider the case where a MN is compromised but not duplicated (e.g., the mobile device in stolen): Well-established mechanisms (e.g., blocking the node's account) can be used in this case.

4.2 Security Analysis

In this subsection, we will analyze the robustness of our protocols against these attacks.

Publicity attack: If S does not provide MN with the promised QoS, MN will send a negative report to TCA. If this attack is repeated, the cumulation of the negative reports will affect the future reputation records of S. If on the contrary, this attack is performed rarely, it will not affect much the reputation of S but S gains almost nothing from performing this attack; as S is rational, it will not perform this attack.

The same reasoning holds if $S=H$ with, in addition, the possibility for MN to punish H by choosing another home WISP.

Selective publicity attack: The anonymity of the mobile nodes prevents S (if $S \neq H$) from performing the Publicity attack against a specific MN. The only possible selection would be based on the home network (i.e., S performs the Publicity attack with all the MNs affiliated with a given home network). S gains nothing from this attack and thus S will not perform it.

Denigration attack: If MN does not send the report on the satisfaction level, H will not give it the ε reward and TCA will consider the absence of feedback as negative feedback. Therefore, this attack is not rational for MN.

So it is more interesting for MN to send a negative feedback instead of not sending the report at all: The effect of the attack is the same and at least MN will get paid for the sending. But this attack is still not rational. Indeed, MN gains nothing from sending a negative feedback instead of a positive one (the cost of the sending remains the same). Such behavior is thus purely malicious.

This attack is not harmful for the WISP, unless it is performed systematically and by a high number of colluding attackers. However, TCA can statistically detect it if the following events happen frequently[7]:

- The MNs affiliated with H always pretend that they received a bad QoS from a given WISP (from a given hot spot managed by that WISP), whereas many other MNs report on a good QoS on that very WISP[8]. As the selective publicity attack is not possible, this situation is suspect and TCA may punish[9] H.

- TCA never receives reports from MNs affiliated with H about the sessions they established with S.

- The MNs affiliated with H pretend that the QoS was bad but at the same time the duration of the session and the amount of data exchanged prove that the QoS was good[10].

Note that this attack comes with an important cost: if an attacker $\mathcal{A}$ wants to alter the reputation of S by parking misbehaving nodes close to the hot spots managed by S, $\mathcal{A}$ should own many devices and devote them to the attack. Note also that this colluding attack may harm very small WISPs (with few number of hot spots) - if the attacker pays the price - but it is much too costly against WISPs with hundreds or thousands hot spots.

Flattering attack: It is not rational for MN to send a positive feedback if it receives a bad QoS from S, unless it has an incentive to do so (e.g., S remunerates MN for the reports).

This attack improves the reputation of the targeted WISP only if it is performed systematically and by a high num-

---

[6]As mentioned in Subsection 2.2, $\mathcal{A}$ can be a mobile node or a WISP.

[7]The higher the number of events is, the more accurate the detection is. Note that statistical detection techniques do not hold if the majority of the nodes are misbehaving, which is not likely to be the case in WiFi networks.

[8]In order to have more accurate detection, TCA can consider each access point of the WISP separately.

[9]TCA, which is a honest and impartial party, can punish H by downgrading it's reputation record.

[10]TCA knows the root w_0 of the hash chain from the contract and knows w_{ell} from the report; it can therefore estimate the mount of data exchanged between MN and S.

ber of colluding attackers. The detection mechanism can be similar to the one proposed for the Denigration attack.

However, a specificity of this attack resides in the fact that H can create "virtual" MNs (i.e., MNs that have an account but are not necessarily real devices), emulate connections with them and make them systematically send positive feedback. This leads to a cost that is much lower than the cost of the Denigration attack but TCA can detect it if (i) the MNs affiliated with H rarely connect to foreign WISPs (or at least much less than average) or if (ii) H is not rewarded for the connections it established with a high number of MNs affiliated with it (if we assume that this information is available to TCA).

Report dropping attack: If S expects a negative feedback, it may want to drop the report on the satisfaction level instead of transmitting it to TCA. But as the absence of feedback counts as negative feedback, this dropping does not help S. Furthermore, the report may be positive: Assuming that the feedback is defined between values *minRep* and *maxRep*, not receiving the report corresponds to a feedback of *minRep*. This attack is therefore not rational.

Service interruption attack: If S refuses to provide the i-th part of the service, MN will keep asking for it (by sending again the i-th payment). After a predefined number of retransmission requests, MN will end the session, which prevents S from providing more service parts (and thus earning more money) and also affects the satisfaction level of MN.

If nevertheless, we want to prevent S from receiving the i-th payment without providing the i-th part of the service, we can use the payment system presented in [4].

Refusal to pay attack: If MN does not send the i-th payment, S will not provide the i-th part of the service and the session will end (after a predefined number of retransmission requests). This attack is then not rational: It prevents MN from receiving the service part but does not harm S.

Repudiation attack: This attack is not possible because H and TCA receive the contract C from both MN and S (Messages 8 and 10). The two copies should be identical, otherwise TCA will not send the message 11 and the session setup will not terminate. Therefore, once the session is established, MN and S cannot retract their agreement.

To prevent S or MN from sending a correct information to TCA but not to H, we can also require a response from H to establish the session.

Packet dropping attack: If a message is not generated or is dropped during session setup, the secure session will not be established. If $\mathcal{A} = MN$ (i.e., MN does not generate messages 1, 3, 7 or 9), it will not be able to connect to the Internet but does not harm S. If $\mathcal{A} = S$, it will not provide the part of the service to MN; MN will select another WISP and S would lose an opportunity for revenue.

If during the secure session, the payment proof or the part of the service is not generated or is dropped, the entity that is waiting for it asks for retransmissions (if needed several times). If it does not receive the message, the session is closed.

If S does not forward the satisfaction level of MN, it is equivalent to the denigration attack (see Subsection 4.2).

If S does not generate the payment request and sends it to H (Message 15), it will not get rewarded for the service parts it provided to MN.

Filtering attack: The messages exchanged between the different parties in our protocols are cryptographically protected, using MAC computations or digital signatures. Therefore, any modification of a message will be detected at the receiver. Therefore, tampering with a message is equivalent to not sending the message at all (an incorrect message is discarded) and it is treated in the same way (see the *Packet dropping* attack).

Replay attack: During session setup, the messages exchanged between the different entities (Messages (2) to (6)) are protected using nonces; the delayed messages are detected and discarded.

During the secure session: the payment proofs and the parts of the service arrive in sequence; a replay is immediately detected and discarded.

During session closing, the payment request and the satisfaction level (Messages (15), (17) and (18)) are expected only once; a replay is immediately detected and discarded.

4.3 Overhead

In this subsection, we evaluate the computation and communication overhead of our solution for a mobile node. We consider only the mobile node because it is the only entity that is severely ressource restrained and because in this way we cover all the wireless communications.

4.3.1 Computation overhead

During the different phases of our protocols, we use symmetric key and public key cryptography primitives to secure the message exchange and to correctly authenticate the different message parties involved in the communication. We minimize however the use of public key cryptography, especially by the mobile nodes, to reduce the computation cost.

Hence, MN uses public key primitives only for two messages: it verifies the certificate, the signature and the reputation of the WISP it selects (Message 2) and it encrypts a message for TCA (Message 9). For all other messages, MN uses symmetric key cryptography primitives: $5 + 2\ell$ MAC operations (ℓ being the total number of service parts), 2 encryptions and 1 decryption.

Public key operations are also used in the message exchange between TCA and the two WISPs S and H (Messages 11, 18 and 19). It is however possible to commute them into symmetric key operations, if we assume that S and TCA establish a symmetric key when they first begin their interaction.

Note that the existence of a tamperproof hardware at MN is not necessary for the good functioning of our protocols, but it may be a good solution for protecting the long term symmetric key k_{HM} that MN shares with H.

4.3.2 Communication overhead

Table 1 provides reasonable values of the size of the different fields appearing in our protocol.

Field Name	ReqID	IDs	n_M,pad	w_i	ℓ
Size (bytes)	4	16	20	20	2
Field Name	MAC	PK	QoS, P, R	k	tag
Size (bytes)	16	150	1	16	6

Table 1: Size of the fields used in our protocol

ReqID is encoded on 4 bytes to reduce the risk of using the same identifier for two different requests. The identifiers of the WISPs and the nodes (W, H, S, MN and TID) are encoded on 16 bytes (assuming e.g. an IPv6 format). The paywords w_i are encoded on 20 bytes (assuming e.g. SHA) and the QoS (AQ and $QoSEval$), the reputation R and the price P are encoded on 1 byte each (which is enough to encode values between 0 and 100). The symmetric keys k_{HM}, k_{HS}, k_{MS} and k_{MT} are encoded on 16 bytes (128 bits) and the public keys are encoded on 150 bytes (assuming e.g. RSA, see [13]). We encode the nonce n_M and the pad on 20 bytes, the *tag* on 6 bytes (see Subsection 3.1.2) and MAC on 16 bytes. Finally, we encode ℓ on 2 bytes to support long sessions.

We consider the example where MN is downloading a 1 MB file. The file is divided into 1 KB packets and each 50 packets represent a part of service ($\ell = 20$ parts of service in total). Using the values of Table 1, an end-to-end session between MN and S represents an overhead, for MN, of 18337 bytes, which represents an overhead per packet of around 18 bytes (i.e., less than 2% of the packet size).

5. REPUTATION SYSTEM ASSESSMENT

Our solution motivates the different players to participate in the reputation mechanisms. Indeed:

- W is motivated to provide MN with the QoS it promised because otherwise the feedback of MN will be negative (see the analysis of the *Publicity* attack in Subsection 4.2).

- MN is motivated to report on its interaction with W because it receives a refund ε.

- W is motivated to forward the report (see the analysis of the *Report dropping* attack in Subsection 4.2).

However, we want also to study the effect of the reputation mechanism on the behavior of the WISPs, i.e., the QoS they effectively offer to the mobile users. We therefore implemented our set of protocols using ns-2 simulator [10].

Using these simulations, we want to verify that:

- The WISPs are encouraged to provide the MNs with a good QoS;

- The WISPs are discouraged from advertising a QoS that is different from the QoS they can really offer;

- It is possible for a WISP that has a bad reputation record to improve its reputation.

5.1 Simulations setup

5.1.1 Decision making at MN

During the WISP selection phase, MN receives several offers from the WISPs. For each offer $Offer_W$, MN computes a value $Decision_W = Rep_W^\alpha \cdot AQ_W^\beta \cdot P_W^{-\gamma}$. It then determines

$$Decision_S = \max_W \{Decision_W\}$$

and selects the WISP S.

- Rep_W is the reputation of the WISP W: It is a value between $minRep=0$ and $maxRep=100$.

- AQ_W is the QoS advertised by W: For the sake of simplicity, we also assume that it is a value between $minQoS=0$ and $maxQoS=100$.

- P_W is the price W is demanding for each part of the service.

- The exponents α, β and γ are parameters that depend on the application MN is running; they are used to emphasize the importance of the variables (Rep_W, AQ_W or P_W). We consider as an example the two following applications:

  - Chat: The user is most likely to choose the WISP that asks for the lowest price. Therefore, we set α, β and γ to 2, 1 and 3, respectively.

  - File transfer: The user is most likely to choose the WISP that offers the highest QoS. Therefore, we set α, β and γ to 2, 2 and 1, respectively.

Note that:

- In order to minimize his personal involvement, the user should set the parameters α, β and γ, for each family of applications, once and for all. However, he should have the possibility to modify them if needed.

- The traffic model (i.e., the frequency at which the packets are sent from S to MN) is the same for the three applications. The only difference is in the choice of the parameters α, β and γ.

- If two (or more) WISPs have the same D_W, one of them is selected at random.

More sophisticated utility functions can include criteria such as the minimum QoS MN is expecting or the maximum price it is willing to pay.

5.1.2 Service provision and QoS

The real QoS RQ_S, $0 \leq RQ_S \leq 100$ received by MN can be different from AQ_S (the QoS advertised by S during session setup). During the implementation of our set of protocols, we represented the behavior of S whose real QoS is RQ_S as follows[11]: Each time S has to provide a "part of service"[12] to MN, it sends it with a probability $RQ_S/100$. If MN does not receive the packet, it sends a retransmission request to S. After 4 unsuccessful retransmission requests, MN closes the session with S. The time during which MN is waiting for the packets and asking for retransmissions represents a delay that justifies the decrease of the QoS offered by S.

5.1.3 Satisfaction level report

At the end of each session, MN evaluates the real QoS it received from S. There can be different levels of satisfaction for this evaluation. We provide here a simple example based on packet counting:

$$RQ_S = max(0, \frac{nbPkts - nbRetReq}{nbPkts} \cdot maxQoS)$$

where $nbPkts$ is the total number of packets it received[13] from S and $nbRetReq$ is the number of retransmissions it had to request.

---

[11] As mentioned in Subsection 5.1.1, we assume that $minQoS=0$ and $maxQoS=100$.

[12] For the sake of simplicity of explanation, we consider in our implementation that the provider sends one part of service per packet.

[13] In the special case where $nbPkts = 0$ (i.e., MN receives no packet from S), we have $RQ_S = 0$.

Then, MN compares RQ_S to AQ_S by computing:

$$QoSEval_{S,CID} = \frac{RQ_S}{AQ_S}$$

5.1.4 Reputation records update

TCA updates the reputation records every 2000 seconds. The new reputation $newRep_S$ of S is computed as follows:

$$newRep_S = \lambda \cdot Rep_S + (1 - \lambda) \cdot \frac{\sum_{CID} QoSEval_{S,CID}}{nbSessions_S}$$

where Rep_S is the current reputation of S, $nbSessions_S$ is the number of sessions established by S (and already closed) during the last 2000 seconds and $feedback_S$ is the sum of all $QoSEval_S$ received over all these sessions (the absence of feedback is considered as $QoSEval_S = 0$). λ represents the "weight of the past" and is set to $1/2$ in our simulations.

Note that if S advertises a QoS that is lower than the real QoS it offers (i.e., $AQ_S < RQ_S$), we will have $QoSEval_S > maxRep$, which may lead to a new reputation that is also higher than $maxRep$. If it is the case, *TCA* keeps $newRep_S$ as it is in its database but sends to S a new reputation record equal to $maxRep$.

5.1.5 Simulation environment

We consider a network of 5 WISPs and 50 MNs. The WISPs are numbered from 1 to 5 and for each WISP, we define the advertised QoS, the real QoS and the price it asks for each part of the service. We initialize the reputation of the WISPs to $maxRep = 100$. MNs and WISPs are static[14] and each WISP is a home WISP for 10 MNs. Each simulation lasts for 50000 seconds and the reputation updates are made every 2000 seconds.

We consider that a WISP W is:

- "honest" if it advertises the real QoS it is offering (i.e., $RQ_W = AQ_W$),

- "misbehaving" if it advertises a QoS that is higher that the real QoS it is offering (i.e., $RQ_W < AQ_W$),

- "modest" if it advertises a QoS that is lower than the real QoS it is offering (i.e., $RQ_W > AQ_W$).

We conducted three sets of simulations to study three aspects of our solution:

Set 1: We want to study the reaction of the network if all the WISPs are honest but offer different QoSs: WISPs 1, 2, 3, 4 and 5 advertise and offer QoS = 60, 70, 80, 90 and 99, respectively[15]. We consider the two following scenarios:

Scenario 1.1: All the WISPs ask for the same price. At the beginning of a simulation, we assign to each MN, with equal probability, one of the two following applications: chat or file transfer (see Subsection 5.1.1).

Scenario 1.2: The WISPs ask for prices that are proportional to their QoSs ($P_W \sim RQ_W$). We expect the choice of the application to have an effect on the results, so we run 2 sets of simulations; one for each kind of application (i.e., all the nodes run that application).

---

[14]All MNs are within the power range of all WISPs, it is therefore useless to consider mobility in this case.

[15]We do not consider the case where $AQ = 100$ because such a perfect case is probably not possible in real life conditions.

Set 2: We want to study the reaction of the network to the presence of misbehaving WISPs and modest WISPs: WISPs 1, 2, 3, 4 and 5 advertise $AQ = 60$, 70, 80, 90 and 99, respectively; but all of them offer $RQ = 80$. We consider the two following scenarios:

Scenario 2.1: All the WISPs ask for the same price. At the beginning of a simulation, we assign to each *MN*, with equal probability, one of the following applications: chat or file transfer.

Scenario 2.2: The WISPs ask for prices that are proportional to their QoSs ($P_W \sim RQ_W$). We expect the choice of the application to have an effect on the results, so we run 2 sets of simulations; one for each kind of application (i.e., all the nodes run that application).

Set 3: We assume that all the WISPs are honest, offer the same QoS and ask for the same price. At the beginning of a simulation, we assign to each *MN*, with equal probability, one of the following applications: chat or file transfer. We want to study the effect of the initial reputation of a WISP that opens its service. We assume that the newcomer is WISP 1 and we consider the three following scenarios:

Scenario 3.1: The initial reputation of WISP 1 equals the one of the other WISPs ($Rep_1 = maxRep = 100$ because the WISPs are honest).

Scenario 3.2: The initial reputation of WISP 1 is lower than the one of the other WISPs ($Rep_1 = 50$).

Scenario 3.3: The initial reputation of WISP 1 is lower than the one of the other WISPs ($Rep_1 = 50$) but WISP 1 asks for a lower price.

5.2 Simulation Results

We run 10 simulations for each of the scenarios listed in Subsection 5.1.5 (i.e., the plots represent the average over the 10 measurements). Each WISP W is characterized by the triplet (AQ_W, RQ_W, P_W) (See the legend in Figures 3 to 11). The results are the following:

Set 1: The results for Scenario 1.1 show that if all the WISPs ask for the same price, almost all the users select the WISP that offers the best QoS (WISP 5 in Figure 3). The other WISPs (mainly WISP 4) can occasionally have some clients because the randomness introduced for the service provision at the WISPs (see Subsection 5.1.2) may lead to a slight decrease in WISP 5's reputation.

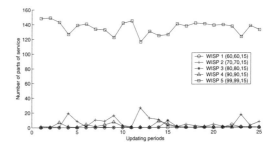

Figure 3: Results for Scenario 1.1.

The results for Scenario 1.2 show that if all the WISPs offer different QoSs and ask for different prices, the choice of the users depends on the application they are running; e.g., if the nodes run a chat application (see Figure 4), the majority of the nodes choose the WISP 2 whereas if the nodes run a file transfer application (see Figure 5), the majority of the nodes choose the WISP 5 that offers the best QoS.

Note that in Scenario 1.2, nodes running the chat application do not choose WISP 1 even if it offers a lower price than WISP 2. By analyzing the data, we realized that this is because the reputation of WISP 2 is significantly higher than the one of WISP 1, which is caused by the randomness introduced, for the service provision, at the WISPs (see Subsection 5.1.2).

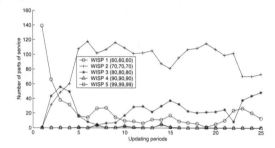

Figure 4: Results for Scenario 1.2 (chat).

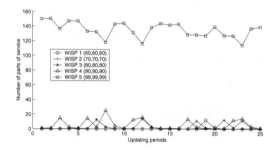

Figure 5: Results for Scenario 1.2 (file transfer).

These results clearly prove that:

- the WISPs are encouraged to provide a good QoS and

- honest WISPs offering different QoSs can co-exist in the same network.

Set 2: The results for Scenario 2.1 show that if all the WISPs ask for the same price, most of the users select the WISP that offers the best real QoS (WISP 3 in Figure 6). Modest WISPs (here WISPs 1 and 2) and misbehaving WISPs (here WISPs 4 and 5) are selected much less often.

Note that the mobile nodes have no direct indication on the real QoS of the WISPs. They are however able to correctly evaluate the behavior of the WISPs because the correspondence between the advertised QoS and the real QoS is taken into consideration in the updating of the reputations.

The results for Scenario 2.2 show that almost all the nodes that run the chat application (see Figure 7) choose WISP 1, which offers the lowest price and at the same time has a very good reputation. The majority of the nodes running a file transfer application (see Figure 8) choose WISP 3 because it offers the best real QoS.

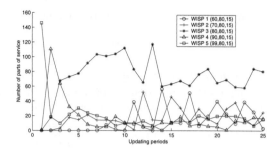

Figure 6: Results for Scenario 2.1.

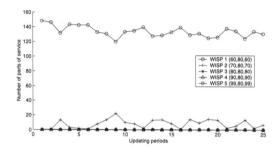

Figure 7: Results for Scenario 2.2 (chat).

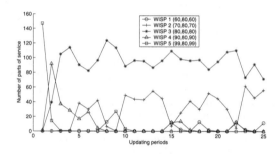

Figure 8: Results for Scenario 2.2 (file transfer).

These results clearly prove that the WISPs are discouraged from misbehaving (i.e., to advertise a QoS that is higher than the real QoS they can offer) and from being modest (i.e., advertising a QoS that is lower than the real QoS they can offer).

Set 3: In Scenarios 3.1 and 3.2, all the WISPs offer the same QoS and ask for the same price.

The results for Scenario 3.1 show that if WISP 1 has, when it opens its service, the same reputation as the other WISPs, it has more or less the same probability to get clients as others do (see Figure 9).

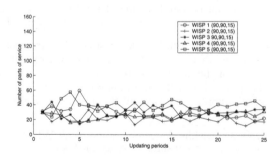

Figure 9: Results for Scenario 3.1.

The results for Scenario 3.2 show if WISP 1 has, when it opens its service, a reputation that is lower than the reputation of all other WISPs, it has no chance to get clients. (see Figure 10).

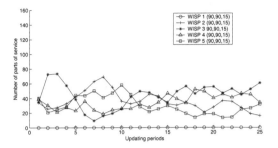

Figure 10: Results for Scenario 3.2.

In Scenario 3.3, all the WISPs offer the same QoS and all of them, except WISP 1, ask for the same price; WISP 1 asks for a much lower price (3 times less than for the others). The results show that by decreasing the price it is asking for, WISP 1 can "reintegrate" the network and get the clients.

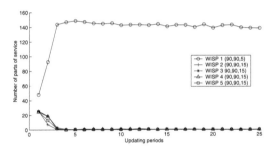

Figure 11: Results for Scenario 3.3.

Note that even if according to the results WISP 1 gets almost all the clients, it is not interesting for it to keep the price very low because it will probably not cover its expenses; lowering the prices can therefore be considered a way of "launching" (if the initial reputation is not $maxRep$) or of "redemption" (if the WISP damaged its own reputation because it misbehaved).

These results clearly prove that:

- the initial reputation of the WISPs should be set to $maxRep$, not to oblige them to lower their prices[16]. If afterwards they do not offer a good QoS or if they misbehave, they will be punished as we showed in the previous scenarios.

- if the reputation of a given WISP decreases because it misbehaves, this WISP is still able to reintegrate the network. However this reintegration comes with a cost (i.e., asking for a price that is much lower than usual).

5.3 Prediction of the QoS offered by the WISP

In Subsection 2.1, we assume that S is able to evaluate the QoS it provides to the mobile nodes; in the simplest

---

[16]A WISP trying to cheat by changing its identity would be detected by the TCA (because it has to register with it each time).

implementation, this QoS would be limited to the mean bitrate; more sophisticated solutions would consider additional parameters such as the provided peak rate, the maximum delay, and the maximum delay jitter; this would be notably the case with IEEE 802.11e [12]. Indeed, the proper operation of our protocols requires S to be able to predict the QoS that it will be able to offer (see the results for the second set of simulations in Subsection 5.2).

To the best of our knowledge, there is no well-established QoS "prediction" technique in CSMA/CA network. We propose the following, statistics-based solution: while it operates, S maintains:

- the history of its connections with the mobile nodes,

- the QoS it was able to offer to them, and

- the conditions under which this QoS provision was possible, such as (i) the number of MNs served simultaneously per hot spot; (ii) the number of neighboring access points (i.e., taking interference into account); (iii) the period of the day (e.g., peak hours, etc.); (iv) the period of the year (e.g., working day, week-end, holidays, etc.).

Using this information, S predicts the QoS it can offer. It can then for example decide to what extent it wants to "overbook" itself. This QoS prediction can be combined to the use of a Differentiated Bandwidth Allocation similar to the one proposed in the CHOICE architecture [1, 15].

6. STATE OF THE ART

Reputation-based systems: These systems are mainly used to build trust and foster cooperation among a given community. The efficiency of reputation mechanisms have been widely studied in various fields and with different approaches. Studies such as [8, 18, 19] consider the effect of *online* reputation systems [5] on e-marketing and trading communities like e-Bay. Reputation mechanisms are also used to foster cooperation in peer-to-peer networks [6] or in ad hoc networks [3, 14].

But, from all these studies, we cannot draw a clear conclusion about the efficiency of reputation systems; each of these mechanisms should thus be analyzed on a per-case basis.

Roaming in WISPs: The deployment and success of WiFi networks is slowed down by the lack of interoperability between WiFi providers (also called *fragmentation* problem [16]): A client that has an account with a WISP A cannot connect to a hotspot managed by a WISP B. However, the situation is changing and more and more WISPs are establishing roaming agreements (similar to what is done for cellular networks). The roaming can be between providers within the same country (e.g., T-Mobile and iPass in the US) or international (e.g., between the British BT and the American *Airpath*).

Another solution would be to use the service of a *WiFi roaming operator* such as *Boingo Wireless* [9]. Such an operator tries to solve the roaming problem by having agreements with as many WISPs as possible. Then it aggregates all the hot spots managed by these WISPs into a single (seamless) network. However, Boingo does not consider the problem of the variable QoS in WiFi networks.

In [17], Patel and Crowcroft propose a ticket based system that allows mobile users to connect to foreign service providers: The user contacts a *ticket server* to acquire a ticket, requests a service from a *service server* and uses the ticket to pay for that service. However, unlike the solution we present in this paper, the authors do not question the honesty of the service providers i.e., they assume that the service providers provide the users with a good quality of service, which is far from being guaranteed in WiFi networks. The same problem exists in the solution proposed by Zhang et al. [22].

In [7], Efstathiou and Polyzos present a Peer-to-Peer Wireless Network Confederation (P2PWNC) where the roaming problem is considered as a peer-to-peer resource sharing problem. Indeed, they propose a solution where a WISP has to allow the foreign users to access its hotspots in order to allow its own users to connect to foreign WISPs' hotspots. However, this solution presents the same problem as for [17], i.e., there is no guarantee of a good QoS provision.

In [2], we considered also the problem of interoperability between the WISPs and we used a reputation system to foster good QoS provision. However, the solution proposed in [2] differs from the one presented in this paper in two main points. The first difference is the trust model: In [2], we consider that even if H is itself a WISP, it plays only the role of a home network and is trusted by all other parties. On the contrary, S is considered as rational (i.e., it can cheat if it is beneficial). We think that this assumption should be relaxed because H can be a home WISP for some nodes but, at the same time, a foreign WISP for other nodes; assuming that it will be rational and honest at the same time makes no sense. The second difference is in the content of the paper: Compared to [2], here we present the details of the protocols, we offer a detailed security analysis of the solution and we evaluate the reputation system.

7. CONCLUSION

The work presented in this paper describes a simple solution that enables a mobile node to connect to a foreign WISP in a secure way while preserving its anonymity and meanwhile discouraging the WISPs from intentionally providing the mobile users with a bad QoS.

We have analyzed the robustness of our solution against different attacks and we have shown that our protocols thwart rational attacks, detect malicious attacks and identify the attacker.

We have proved by means of simulations that the WISPs are encouraged to provide the MNs with a good QoS and, at the same time, discouraged from advertising a QoS that is different from the QoS they can really offer.

In terms of future work, we plan to study more in detail the prediction of the QoS the WISPs can offer to their clients and the cheating detection techniques. We also plan to investigate the feasibility of a "multi-hop WiFi network" (i.e., a WiFi network that is extended using multi-hop communications [21]) in terms of network performance and security.

8. REFERENCES

[1] P. Bahl, A. Balachandran, A. Miu, W. Russell, G. Voelker, and Y.M. Wang. PAWNs: Satisfying the Need for Ubiquitous Connectivity and Location Services. *IEEE Personal Communications Magazine*, 9(1), 2002.

[2] N. Ben Salem, J.-P. Hubaux, and M. Jakobsson. Fuelling WiFi deployment: A reputation-based solution. In *Proceedings of WiOpt*, 2004.

[3] S. Buchegger and J.-Y. Le Boudec. Performance Analysis of the CONFIDANT Protocol: Cooperation Of Nodes - Fairness In Distributed Ad Hoc NeTworks. In *Proceedings of MobiHOC*, 2002.

[4] L. Buttyán. Removing the Financial Incentive to Cheat in Micropayment Schemes. *IEE Electronics Letters*, January 2000.

[5] C. Dellacrocas and P. Resnick. Online Reputation Mechanisms - A Roadmap for Future Research. In *1st Interdisciplinary Symposium on Online Reputation Mechanism*, 2003.

[6] Z. Despotovic and K. Aberer. Trust and Reputation in P2P networks. In *1st Interdisciplinary Symposium on Online Reputation Mechanism*, 2003.

[7] E.C. Efstathiou and G.C. Polyzos. A Peer-to-Peer Approach to Wireless LAN Roaming. In *Proceedings of WMASH*, 2003.

[8] D. Houser and J. Wooders. Reputation in Auctions: Theory, and Evidence from eBay. Working Paper 00-01, University of Arizona, 2001.

[9] http://www.boingo.com/.

[10] http://www.isi.edu/nsnam/ns/.

[11] http://www.rsasecurity.com/products/securid/.

[12] IEEE 802.11 WG, Draft Supplement to Standard for Telecommunications and Information Exchange between Systems-LAN/MAN Specific Requirements- Part 11: Wireless LAN MAC and Physical Layer (PHY) Specifications: Medium Access Control (MAC), Enhancements for QoS, 802.11e Draft 4.1, February 2003.

[13] A. K. Lenstra and E. R. Verheul. Selecting Cryptographic Key Sizes. *Journal of Cryptology: the journal of the International Association for Cryptologic Research*, 14(4), 2001.

[14] P. Michiardi and R.Molva. Core: A Collaborative Reputation Mechanism To Enforce Node Cooperation In Mobile AD HOC Networks. In *Proceedings of The 6th IFIP Communications and Multimedia Security Conference*, 2002.

[15] A. Miu and P. Bahl. Dynamic Host Configuration for Managing Mobility between Public and Private Networks. In *The 3rd Usenix Internet Technical Symposium*, 2001.

[16] Boingo Wi-Fi Industry White Paper. Towards Ubiquitous Wireless Broadband. http://www.boingo.com/wi-fi_industry_basics.pdf, 2003.

[17] B. Patel and J. Crowcroft. Ticket based Service Access for the Mobile User. In *Proceedings of MobiCom*, 1997.

[18] P. Resnick and R. Zeckhauser. Trust Among Strangers in Internet Transactions: Empirical Analysis of eBay's Reputation System. In *NBER workshop on empirical studies of electronic commerce*, 2001.

[19] P. Resnick, R. Zeckhauser, J. Swanson, and K. Lockwood. The Value of Reputation on eBay: A Controlled Experiment. In *ESA Conference*, 2002.

[20] R. Rivest and A. Shamir. PayWord and MicroMint: Two simple micro-payment schemes. Technical report, MIT Laboratory for Computer Science, 1996.

[21] N. Ben Salem and M. Jakobsson L. Buttyan, J.P. Hubaux. A charging and rewarding scheme for packet forwarding. In *Proceedings of MobiHoc*, 2003.

[22] J. Zhang, J. Li, S. Weinstein, and N. Tu. Virtual Operator Based AAA in Wireless LAN Hot Spots with Ad Hoc Networking Support. *Mobile Computing and Communications Review*, 6(13), 2002.

Smart Edge Server – Beyond a Wireless Access Point

G. Manjunath, T. Simunic, V. Krishnan,

J. Tourrilhes, D. Das, V. Srinivasmurthy, A. McReynolds

Hewlett Packard Labs

1501 Page Mill Rd.

Palo Alto, CA 94304

{geetham, ddas,venuks}@india.hp.com {tajana, venky, jt, allanm}@hpl.hp.com

ABSTRACT

Wireless access at cafes, airports, homes and businesses have proliferated all over the globe with several different Wireless Internet Service Providers. Similarly, digital media has created a paradigm shift in media processing resulting in a complete change in media usage models, revamped existing businesses and has introduced new industry players. We believe there is a tremendous opportunity for application and system services at the intersection of the above two domains for exploiting the wireless connectivity to provide ease of use in handling media. In this paper, we propose a feature-rich, secure wireless service delivery framework over enhanced public access points (called Smart Edge Servers), which provides the right platform for deployment of specialized services to the mobile users. The Smart Edge Server provides secure wireless access to the clients, has sophisticated media handling and storage capabilities and uses advanced techniques to manage resources available to it, such as bandwidth, power and the type of connectivity. A prototype implementation of our Smart Edge Server has been built that implements all the features discussed above.

Categories and Subject Descriptors

C.2.5 [**Computer-communication networks**]: Local and Wide-Area Networks – *Internet, Wireless.*

General Terms

Management, Measurement, Performance, Design, Security.

Keywords

access point, management, media, security, wireless, low-power

1. INTRODUCTION

Wireless network access points, known as hotspots, are proving to be a cost effective and viable solution for ubiquitous access to the Internet [1]. This has led to an explosive growth of wireless access points at cafes, airports, homes and offices. Typically wireless service vendors offer only basic connectivity to the Internet with minimum provision for secure communication and no provision for management of resources, such as battery lifetime on the mobile device. Commercial wireless gateways available from Orinico, BlueSocket, and NetMotion provide only secure physical access to the Internet. The HP ProCurve series of wireless access point products provides campus-wide security with support for seamless roaming across the physical network and accommodates precise network access control. The HP Wireless Connection Manager, used in the popular alliance with Starbucks, is a free software application that automatically detects, connects, and facilitates user mobility across different high-speed wireless networks. A number of researchers have proposed various methods to seamlessly handoff between the wireless networks at both macro and micro level [19], [20], but most have not taken mobile's power limitations and QoS needs of multimedia into account.

Concurrently there has been a paradigm shift in multimedia processing caused by digitized media. For example, the digital camera experience is very different from the traditional film camera – there are no consumables such as film rolls, job of film processing shifted to the owner of the camera, etc. In general, when it comes to handling media, the user's expectations are much like that for just another appliance – the digital media devices should just work. In addition, there has been record growth in media downloading and sharing [18]. There is more and more demand for secure and easy media access, with a significant fraction of it being personal media that is shared. This has resulted in introduction of new industry players, revamping of existing businesses and a complete change in usage models.

There is a tremendous opportunity for application and system services at the intersection of the above two trends. Wireless Access points, by leveraging their placement at the edge of a content delivery network, are very suitable for deploying content-based, customized services and enable special system services for the mobile client. This niche has not yet been captured by any service provider due to lack of an appropriate software infrastructure over these wireless access points. In addition, because access points deliver content to multiple clients in their environment over potentially a number of different wireless standards (e.g. WLAN, Bluetooth, GPRS), they are in the best position to carefully manage client's resources such as battery lifetime and mobility.

In this work we present a secure wireless service delivery framework over enhanced wireless access points which provides the right platform for deployment of specialized services and applications to mobile users. We call these enhanced, feature-rich access points Smart Edge Servers (SeS). An SeS has several new features that are above and beyond what today's wireless access points provide, such as content adaptation, energy saving, seamless migration across different wireless links, consolidated access to distributed personal media and secure access to services.

2. SES OVERVIEW

Our solution consists of two parts namely, the server-side and the client-side. Smart Edge Server, the server-side module, is an enhanced wireless access point whose components are depicted pictorially in Figure 1. The Client side resides on a mobile device (e.g. PDA) and is a proxy with security related enhancements and resource management hooks. The SeS is further organized into three major management components: security, media and resource manager. Security manager handles all issues related to authentication and secure communication between the client and the SeS. Media manager performs a wide variety of tasks, ranging from media content adaptation, to virtualization of media storage. Resource manager is capable of delivering a good quality of service to a client while increasing battery lifetime and seamlessly migrating wireless connection between different wireless network interfaces. Specific functions of these components are elaborated in the subsequent sections. In this paper we present a prototype implementation of the SeS. Although we use browser (HTTP protocol) interaction with a user as a sample scenario throughout the paper, our work is completely applicable and has been used with other applications and protocols.

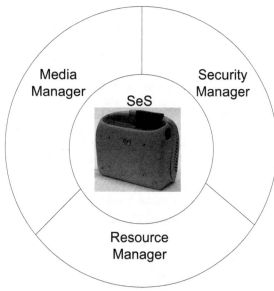

Figure 1 Main components of SeS

3. SECURITY MANAGER

There are many different Wireless ISPs deploying access point solutions at public locations [2]. Just as in any public wireless network, a mobile user gets network and local services by subscribing to the ISP deploying the SeS. SeS provides user authentication and authorization in such a completely un-trusted environment (public place) that has support for service level agreements among the different service providers so that the user can access services to which he has subscribed elsewhere. The primary responsibility of SeS is to verify the authenticity of the authorizations issued by the service providers prior to allowing the users access to services. This section gives an overview of the SeS security infrastructure that satisfies the above needs.

There are three critical steps which need to be occur before secure data communication can start: registration of SeS and client with internet service provider (ISP), three party authentication (client, SeS and ISP), and the authorization for services requested by the client. Each of these steps is described in more detail below. Although in all the examples we describe our prototype implementation which uses HTTP, our security management infrastructure can operate just as well when using other approaches. For example, protocols such as PANA [16] are used for network authentication. We have also used our key management protocol to set the IPSec [24] secure associations between the client and the SeS for IP level connectivity – enabling secure multimedia and VoIP applications. In addition, SeS facilitates end-to-end secure association between the client and a remote server by acting as a VPN gateway.

In the first step of managing secure communication between the client and the SeS users are assigned a user-id and a password by a Verification Server (VS), a global entity trusted by ISPs. Any service provided by ISP, including connectivity, is further authorized by the ISP's Authorization Server (AS) to which a user must register in order to use that service. The communication between the user, AS and VS is secured using SSL after appropriate certificate validations. Client first submits <user-id, password, VS> securely to the AS of the ISP. The AS then verifies the data with the VS and responds with a generated <shared key, client-id> tuple to the user. This tuple is used for client authentication with the SeS. Our new shared-key distribution technique can be used while the client is mobile and as such provides the following benefits:

- The user-id/password provided to the user by an ISP are used minimally - only for shared key generation, as opposed to it being used every time the user authenticates to an SeS. As a result, the possibility of compromise is curtailed.

- If a device gets lost, the user requests of AS to disable access to just that one device. All other devices in the user's possession can still continue to be used without any reconfiguration since they share different keys with the same AS.

- A policy can be set at the AS whereby a shared key expires after certain duration of time. Using our scheme the shared keys can be generated again and distributed - that will foil any malicious attempt to guess a shared key by analyzing previously captured traffic between the client and AS at the time of authentication and impersonate the user.

- The dynamic generation of <shared key, client-id> provides some anonymity since user names are never disclosed to the SeS - what the SeS sees is the just the client-id..

The next step represents the core of our security infrastructure – our new three-party Key Distribution Protocol [6] where

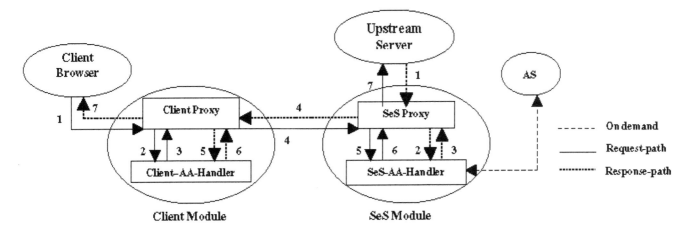

Figure 2: SeS Security Infrastructure

fundamentally two mutually distrusted entities, client and SeS, come together through AS and interact wirelessly. A detailed proof of the correctness of our 3-party key distribution protocol based on BAN logic [23] is provided in [22]. Our three-party authentication protocol accomplishes this critical step in two phases, the *authentication phase* and the *access setup phase*. In the authentication phase, the client, SeS and the ISP (AS server) mutually authenticate each other using the shared key obtained during registration and derive a temporary session key. In the access setup phase, the AS sends the client authorization information to the SeS and securely communicates a service access key to both the SeS and the client. At this point both encryption and authentication keys are generated and exchanged in the a channel secured with the service access key. There is a number of advantages of our three party authentication. Explicit and independent authentication between AS-SeS and AS-Client allows SeS to not have to authenticate with AS for each and every new client. In addition, there is minimal exchange of session keys - limiting the possibility of session key compromise and the possible replay attacks. The session keys are periodically refreshed when the client re-authentications occurs. With our protocol we aim to remove the service-theft and man-in-the-middle attack issues poised by unauthorized clients and rogue SeSs.

As a last step prior to data communication, our security manager sets up the authorization for various services between the client, SeS and the ISP. At the AS, in order to enforce access control for services, the users are classified into realms based on the <user-id, VS> registered with the AS. This service authorization information is also communicated to the SeS during the key distribution protocol for local decision-making. When a user accesses a service owned by an AS, AS_i, the client software initiates a user authentication with AS_i transparently. The authentication and authorization is completely implicit and the security transformations are handled by the server-side and client-side software. Neither the client applications nor the application server need to be aware of the authentication and authorization related details. Our security infrastructure also supports establishment of trust when a user requests a service that is spread across multiple SeS, potentially in different administrative domains [5].

Figure 2 outlines sequence of communication between the client and the SeS through our security management layer here represented by client and server proxy & handlers. Three paths are represented: on-demand path through which both client and the SeS can periodically re-authenticate with the AS, the request path that leads from client to SeS, and the response path from SeS to the client. Here we outline the seven steps illustrated in Figure 2 that are needed for the secure data transmission along the request path (mirrored seven steps are also shown for the response path):

1. The user's HTTP-Client-Software creates a HTTP request and hands it over to the Client Proxy.

2. The Client Proxy accepts the HTTP request and upon receiving, hands it over to the client's authentication and authorization handler (Client-AA-Handler).

3. The Client-AA-Handler, optionally, encrypts the HTTP request using the encryption key. It then computes a Message Authentication Code (MAC) of the request using the authentication key. The MAC, applicable client-id, and the HTTP request are returned to the Client Proxy. The Client-AA-Handler also initiates the protocol described in [6] in case if the session-key is not there or has expired.

4. The Client Proxy frames an additional HTTP header (called AUTH_SPEC_HDR) containing the signature and the client-id. The complete request is then forwarded to the SeS Proxy.

5. The SeS Proxy accepts the HTTP request, and passes it on to the SeS' authentication and authorization handler (SeS-AA-Handler).

6. The SeS-AA-Handler parses the AUTH_SPEC_HDR HTTP header. It uses the client-id as a key to the key table to retrieve the encryption/authentication keys for this client-id and applies reverse transformations on the data. The complete request is then returned to the SeS Proxy along with the authentication and authorization status of the client.

7. The SeS Proxy checks the authentication status of the request, and, if authentic, does the normal HTTP proxying of the request; else, returns back error response to the Client Proxy.

In this section we have outlined how security manager sets up safe communication link between a client and the SeS. The next section presents how SeS manages transfer of media content with a client once the secure connection is established via our security manager.

4. MEDIA MANAGER

SeS views media as a single entity, part of a collection, that is either streamed or is accessed as a chunk of data. Many small appliances sold today have wireless interfaces (e.g. cameras have Bluetooth) whose usage is limited to synchronization with a nearby PC. Our approach greatly increases the potential for data sharing and media accessibility because SeS acts as a secure bridge through which a mobile user can upload data/image/video onto the personal storage without wires. SeS media manager has a number of capabilities, it can:

- *Adapt content:* Each mobile device has a different set of capabilities, thus any media content has to be adapted in order to be presented in a best way. This adaptation is done on the SeS as it has many more resources than a typical client.

- *Interact:* The users can access services that are either local (to the SeS) or remote. SeS supports a browser-based model for interaction from a client, typically a laptop or a handheld.

- *Organize:* Users view their media as a set of collections (as described in the previous section). They can organize their collections by either using tools provided by SeS or by third party solutions operated over the Web folders interface provided by the virtual store.

- *Upload:* Input-generating appliances like cameras can upload their media to the user's repository. The appliance contains a meta-data file that holds keys and other descriptors that provide the SeS information on where the media is to be placed. We developed the notion of *casual-download*, seamlessly uploading pictures from a media device when an SeS is discovered in its immediate neighborhood. The appliances are not always network enabled but have reasonable storage capabilities. This storage is viewed as a cache and when the appliance comes 'within range' of a SeS, the media can get uploaded. SeS provides support for BlueTooth and dock-based devices in addition to flashcard-based uploads.

- *Experience:* For a given media, the SeS generates meta-data descriptors (using MPV) that allow MPV-aware appliances to handle the media. Thus, SeS enables media-centric experiences for a wide range of appliances such as TV, audio players, and laptops.

- *Share:* Users can print photographs and create picture albums on CD-ROMs as SeS enables physical sharing of media too.

We next describe the content adaptation and distributed personal media capabilities (enabling interaction, organization, uploading, and sharing).

4.1 Content Adaptation

Mobile devices are typically limited in computation power, display size and even application capabilities. Much of the content that is available on the wired network is targeted at larger devices, and thus needs to be adapted for better viewing on small client devices. Adaptation of web content can be done potentially in one of the three points of a content delivery network - the client side, the content server or the edge server (SeS). It is not feasible to perform the content adaptation on client devices (e.g. video scaling on PDA) due to their limited computing power and memory; while server side adaptation has practical problems due to the vast number of content providers and diversity in client devices. On the other hand, there are many advantages of edge-side content adaptation. Availability of cached content (before and after adaptation) for use by more than one client helps in faster delivery of the content. The content at the SeS is amenable to adaptation based on the local physical context and network characteristics. Further, specialized machines for specific types of filtering (virus scanning, video transcoding) can be connected to the SeS to perform load balanced filtering under heavy traffic. The rule-based online content adaptation infrastructure at the SeS provides a framework for deploying content adapting services – making the content viewable on the diverse devices and also provides advertising opportunities to the ISP.

Our content adaptation framework is based on the IETF- OPES model supporting the ICAP[3] protocol, an RFC for encapsulating HTTP messages. One of the core components of SeS, the HTTP proxy, is enhanced with an ICAP client that optionally encapsulates all HTTP requests and responses and sends them over for potential modification to an ICAP server (either local or remote). The ICAP server then consults an ICAP rule engine to determine the appropriate filters to be applied on the request/response and adapts accordingly. Typically the SeS administrator configures the rules for adaptation – based on the user policy, QoS requirements, availability of local context dependent services, ISP needs and any specific agreement with a content provider. Figure 3 shows the different active entities of our content adaptation framework in a lifecycle of a request and the longest path a request could take before the response finally reaches the client. The SeS Client (SC) component is configured as the browser's proxy and so initially directs all HTTP requests to the SeS proxy. Additional HTTP headers are used to communicate the client information such as the type of OS and browser specific details.

Various content adaptors (proxylets) installed at the ICAP server may either perform the filtering themselves or utilize a web resource elsewhere to perform the modification. The rule engine performs a rule-based decision to determine which filter(s) to apply, based on the attributes of the request/response. Every rule specifies a regular expression that should match against the value of a HTTP header, and a particular proxylet as the action. The rules are typically set by the SeS administrator using a web-based control provided for remote management. New proxylets can be authored and installed using the same web-based configuration module.

In addition, a set of well-defined proxylet API has been defined to simplify the authoring of proxylets. The rule engine is also integrated with the local Authorization server wherein the SeS administrator would have set access controls for certain filters.

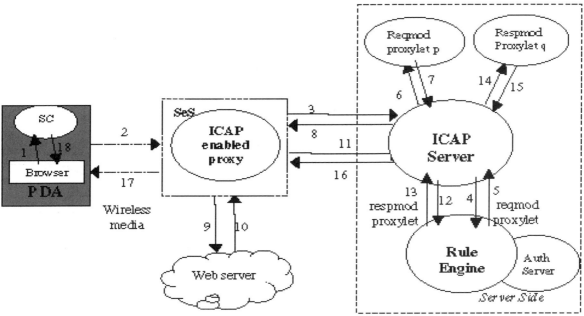

Figure 3: The Online Content Adaptation Framework

Based on the credentials of the client accessing the Server and the filter permissions, the rule engine sends the list of the filters to be enabled for a specific request.

Our infrastructure also supports automatic discovery of available filtering services and applies them based on user preferences. There is also a provision for service level agreements between the filter service provider and the access provider to provide integrated accounting information for a roaming user[13].

4.2 Distributed Personal Media

Distributed Personal Media Service (DPM) is an example application service built on top of the SeS modules. Typically a mobile user has his data on several different devices - home PC, office workstation, laptop, PDA, Camera etc. Ease of access, sharing, synchronization and archival of this distributed data ensuring confidentiality is a real concern. The virtualized storage of SeS enables a consolidated view of all the different pieces of the user's personal media. Further, since the naïve user is comfortable viewing the data as a collection of media as opposed to a directory of files, DPM provides a media collection abstraction of our virtualized storage.

The core component of the DPM is a Storage Virtualizer. We consider each unit of personal media (on different devices) as a "Physical Repository (PR)" and provide a virtual consolidated media repository, called as "Virtual Repository (VR)", to the user. The physical repository could either be a local file system, a network accessible remote file system, a full-fledged storage appliance, disk space bought over the Internet or yet another Virtual Repository. A Virtual Repository is created for every new session initiated by the user and includes the Physical Repositories currently configured in the user's policy. Multiple PR file system interfaces will essentially be supported by the VR – the currently supported ones include standard POSIX file system and WebDAV protocol.

A VR is created on the Smart Edge Server (SeS) closest to the mobile user since the caching and replication algorithms to improve the availability, performance of this storage are best deployed at the edge of a content delivery network. Even the VR provides multiple interfaces for the mobile user – the standard WebDAV protocol, content specific web views through a web service, disk drive for Windows and a mountable file system for Unix clients. The user views a single consolidated store of his media without being aware from which of his personal computing devices the data is being fetched. Any local modification on the personal data is reflected back at the (probably) remote site automatically – the policy for writing back and ensuring consistency is based on the caching/replication policy in effect on the SeS.

Further, we have integrated our security infrastructure described in an earlier section with our virtualized storage to provide user specific views of the storage. Authorization controls can be exercised either at the PR or at the VR. At the PR, WebDAV server is modified to verify the user access permission with the authorization server and formulate an appropriately masked reply as a response to the PROPFIND method of WebDAV. On the other hand, the SeS administrator typically performs the access controls for the VR.

The set of physical repositories to be included in a user's virtualized storage is dependent on the pre-defined policy information that can be specified in multiple forms. The user can specify the network address of his home SeS (also called a Family Data Centre) and DPM would pick up the list of PR's from there. Alternatively, device specific views can be provided through dynamic upload of policy information when the user is first authenticated. A session manager on the SeS manages the user sessions accommodating dynamic *import* of PRs.

A user typically collects all his digitized pictures, audio and video files as albums or collections for ease of access and sharing. Our Distributed Personal Media (DPM), an application over the virtualized storage, provides a collection/album view of personal media – as opposed to a set of media files. This service is typically intended to access and share personal media. Sharing of personal media with friends and relatives needs to be regulated and that regulation should not be very visible. This is ensured in a DPM. For example, if the data viewed is a photo album, a friend who needs enforcement of authentication controls would get a view of the personal media with unauthorized albums masked out. He will be completely unaware of the existence of the unauthorized album. In essence, this results in different views of the virtualized storage for different users. The authorization controls can be exercised at the level of collections using a Web Interface by the owner of the media. The owner can essentially define user-groups (friends, relatives, owner and so on) and allow sharing of the collections for specific groups only.

The collection information is available as a MPV (multi-photo video standard)[17] file. The grouping of media files to form specific collections could therefore be either logical (metadata based, content type, explicit grouping) or physical (based on the their location in the file system). MPV is also used to provide a configurable presentation of a collection based on the media type.

Another interesting view of DPM is as a WebDAV folder as the virtual repository itself is served over the WebDAV protocol. Here, the DPM can be thought of as alleviating the disk space limitation of a client device by providing a consolidated disk drive consisting of all the user's data. If the user creates a file on this disk drive, it would physically reside on a remote machine but provide a local access mechanism to help local applications work on the remote data without any explicit configuration. We also envisage a usage wherein some temporary store is leased out of the local disk space of the SeS and periodically backed up at a remote site (determined by the user policy information).

5. RESOURCE MANAGER
Main resources that SeS manages are client connectivity, power and Quality of Service. As today's clients have multiple different wireless network interfaces (WNICs), the SeS makes decision on when a device should move from one to the other WNIC depending on QoS and power demands. In this section we first describe the technique we used to enable seamless migration from one WNIC to other. We follow up with a detailed description of the policy that decides when migration should occur, and when the currently used WNIC should be in low power state.

5.1 Seamless Wireless Migration
Today's mobile devices support multiple wireless interfaces. Different links offer diverse characteristics in terms of range, speed and power consumption. As new wireless technologies are developed, they will be added to the SeS while the existing link layers can be kept for compatibility with older clients. For the same reasons, the client is also likely to include multiple wireless link types. Thus, the client and the SeS will often have a number of wireless links in common. Depending upon network/device conditions, application and user needs, the best link may be change during a communication session. Our Connection Diversity (CD) framework provides a link level abstraction for seamless connectivity across the diverse physical networks.

A key feature of CD mobility support is that it does not require any support in the infrastructure. This makes it easy to deploy and it enables inter-domain mobility (mobility across different ISPs). Further, our CD framework maintains the same session across multiple links if the interface switch occurs within a single SeS

without any application level support. The CD's name resolver allows the client to interact with the SeS without having to know its DNS name or IP address. It also enables the client to discover other clients and refer to them with a short local name. Lastly, the CD defines API that enable applications on the SeS to get information about specific clients, to know which wireless link they are currently using, and to get event notifications when this changes.

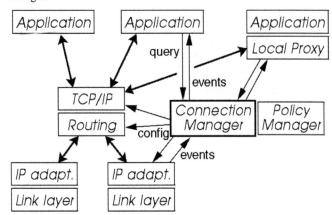

Figure 4: The Connection Diversity Framework

The CD framework shown in Figure 4 is a set of components and interfaces in the client and the SeS, that abstract the various wireless link interfaces and simplify their use. The SeS provides the clients with proper IP configuration via DHCP, runs the connection manager (CM) to monitor and handoff connections, and optionally provides HTTP proxy autoconfiguration. Each client is handled individually, and multiple clients can be supported on the same interface simultaneously. In our current prototype, the SeS keeps all of its interfaces active at all times.

The central piece of the framework is the Connection Manager (CM). The role of the CM is to discover, evaluate, setup and monitor the various paths between the client and the SeS on behalf of the various applications. It directly manages the various link interfaces and includes abstraction modules specific to each link layer used for tight integration with each link interface. The CM performs link discovery to find which paths are available, activates it and configures link interfaces on-demand to enable their use, monitor them for failure, and disconnect them when idle. The CM try as much as possible to use link specific methods for those tasks, for example over BlueTooth it uses the link native discovery, and over 802.11 it uses packet drop events to detect connection failures. It also use generic methods as backup, such as monitoring incoming and outgoing IP packets to detect link idleness and failures. The Policy Manager (PM) component selects the most appropriate link to connect from the client to the SeS based on the current policy, applications requirements and link availability. The CM currently can manage IEEE 802.11b, BlueTooth BNEP and IrDA links. The client component is also integrated with other long-range wireless protocols such as GPRS.

The CD framework provides mobility support to migrate client connectivity seamlessly not only between link layers but also between two SeSs. Mobility between two SeSs is handled using an application layer based hand-off protocol [21] implemented in the Connection Manager of the client. The application uses a

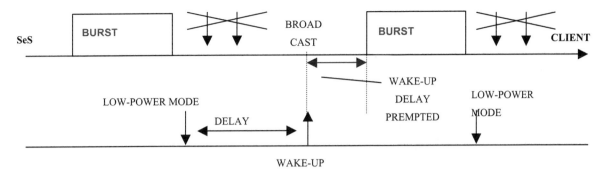

Figure 5: Communication between the client and the SeS for better QoS

direct connection to the Internet, and interacts with the CM to detect connectivity changes and adapts to them. Mobility between two interfaces of the same SeS is handled using a simple vertical handoff protocol [11] implemented in the CM of both the client and the SeS. It switches IP routing between interfaces on both sides of the wireless links – client and SeS.

There is many differences between our approach and traditional mobility protocols such as Mobile IP. Our solution doesn't require any infrastructure support (no Home Agent) and works through NAT and firewall, which significantly eases deployment and allows inter-domain mobility (mobility across different ISPs). Our solution also makes the client local, so it can take advantage of the locality aspects of the SeS, such as a local cache and location aware services. The network performance of our handoff protocol is typically better than MobileIP because the connection is always direct and not through a home agent.

5.2 QoS and Power Management

We have exploited the above feature of wireless migration to provide better client resource utilization. Mobile clients typically have limited battery-lifetime and communication abilities. Our measurements indicate that a large fraction of battery energy is spent for communication (as much as 50%), with more than 90% of that spent listening for any incoming traffic. Thus careful scheduling of communication, and management of multiple wireless interfaces can offer large improvements in battery lifetime of the client device with no perceived performance degradation. We believe that SeSs are great candidates for efficient scheduling as they are not power constrained, and know both wired and wireless network conditions. The SeS obtains the client device characteristics, and monitors communication patterns [8]. Based on this, it seamlessly switches to the appropriate wireless interface on each client, schedules communication sessions and directs when the client enters a low-power state.

The goal of Resource Management (RM) is to enhance the Quality of Service (QoS) while maximizing the battery lifetime of the client devices. RM's primary task is to determine what network interface is most suitable for the client's needs and how to manage its power and performance states. When an application starts on a portable device, RM pre-selects those Wireless Network Interface Cards (WNIC) for data communication whose average throughput is greater than the data consumption rate of the application. As conditions on SeS, wireless link and the client change, the SeS can choose to switch to another WNIC, and also

adapt the times when data is transferred between SeS and the client. This ensures that the QoS requirements of the client's applications are satisfied.

Figure 5 shows how communication occurs between the SeS and the client device. Data is transferred between the client and the SeS in large bursts and then client's WNIC transitions in a low-power mode until the next agreed point of communication with the SeS. During this time much energy can be saved thus enabling longer battery life on the client. The client wakes up on time to receive the next burst of data and thus does not cause any degradation of service. In addition, contention for the wireless medium is now reduced so that SeS is able to support more clients simultaneously. Thus, this efficient control and scheduling of transmission increases the battery life of the client device, increases the accessibility of SeS and improves the quality of service in multiple client environments. The control-related handshaking between the client and the SeS occurs with each data burst.

SeS has to communicate to the client ahead of time how large data burst to receive before going into a low power mode and when to wake up for the next burst. The size of the buffer directly affects the energy spent in communication. If the size of the buffer increases, the average power dissipation of the communication device diminishes due to longer sleep periods and thus less overhead in transition between power states. RM pre-selects WNICs for a particular application based upon their average throughputs and the data consumption rate of the application. The changes in the rates are observed using maximum likelihood estimator. The WNIC that offers minimum power dissipation with regards to communication and RAM is selected. RM also defines the appropriate low-power state of the WNIC along with the switching points. Additionally, it can dynamically switch the selected WNIC if a change in its throughput and/or the average data consumption rate of the application is detected.

6. SES PROTOTYPE

A prototype of our Smart Edge Server, based on off-the-shelf embedded Linux box and our proxy-based software including all of the modules described in the previous sections has been demonstrated within HP. All the local services residing on the SeS are built using Coolbase Web Application server that supports authoring web dynamic services either as C or Python classes [15].

Figure 6: Demo scenario

Figure 6 shows our demo setup. SeS is in the middle with various clients surrounding it: speakers for streaming audio, laptop showing secure access to media storage, Bluetooth camera uploading data to SeS, IPAQ playing back a movie clip and printer with a printed photo. The SeS management infrastructure is designed so that better joint decision making between the three managers (resource, media, security) is possible. In this way the the managers act in concert when it comes to deciding when and over what link to send video frames (resource manager), what size video should be sent (media manager) and what level of encryption should be used (security manager).

As a part of our demo we show an authenticated and authorized mobile user accessing the Internet securely through SeS – using either Wavelan/802.11b or Bluetooth (IPAQ playing video, speakers playing music and laptop showing photos in Figure 6). The content adaptation rules are configured so that MPEG video delivered to the client device is scaled down to fit the size of the screen (Shrek on IPAQ in Figure 6). We also demo a tailored HP PhotoSmart 812 digital camera uploading a captured picture over Bluetooth to a selected album on user's Distributed Personal Media when the camera is in the vicinity of our SeS (camera is next to SeS, the pictures are displayed on the laptop and one is printed on the printer). Similarly, we have shown easy transfer of pictures acquired by Nokia 3650 cell phone through Bluetooth. We next discuss the performance attributes of each component in turn.

Security manager's main contributions are in the area of key distribution, authentication protocol for setting up secure sessions, and authorization techniques. The algorithms for authenticating data communication between the SeS and client and encryption are also used by other standard security protocols, and thus their performance is comparable to what is currently out in the market. Our three-party Key Distribution Protocol has two areas in which very minor performance overhead occurs:

1. The authentication process involves a few protocol messages as described in [6]. This overhead is incurred infreuently and as a result can be neglected.

2. The authorization process involves querying the SeS-AA-Handler (solid lines 5 & 6 as shown in Figure 2) for

each URL requested by the client (in case of HTTP). The overhead here is a call between the SeS Proxy and the AA-Handler. Some of this overhead can be mitigated by caching information at the proxy.

Table 1: Results showing the effect of Content Adaptation

Transfer Characteristic	No Content Adaptation	With Content Adaptation
Movie Size [bytes]	37589390	19484040
Requests per second	0.16	0.31
Time per request: [ms]	6206.488	3233.952
Transfer rate [Kbytes/s]	591.45	588.41
Mean Connection Times [ms]		
Connect:	9	4
Processing:	6196	3229
Waiting:	15	46
Total:	6205	3233

We next compare the performance of our system with and without the content adaptation framework that is a critical part of media manager. A major part of content adaptation is transcoding, which typically results in significantly smaller overall media size transferred. Results in Table 1 show that with our framework we are able to shorten the connectivity time to wireless by more than a factor of two, which also gives a significant reduction in client's energy consumption and an improvement in the overall wireless bandwidth available to the other clients communicating with the SeS.

Table 2: Typical handoff characteristics

Link layer latencies	IrNet	802.11b	BlueTooth
Discovery period	3s	10s	60s
Connection setup	0.8s	0.3s	0.8s
Link breakage detection	1s	0.1s	0.5s

Seamless wireless migration performance is mostly governed by the characteristics of the individual link layers and latency of the events triggering handoff. The values outlined in Tables 2 and 3 are typical of our implementation. The handoff time between link layers of the same SeS is exactly the sum of the link breakage detection time on the old link and connection setup time on the new link [11]. Handoff between two SeS takes more time ; Table 3 shows the breakdown of a handoff from BlueTooth to 802.11b between two SeS [21]. Clearly the extra buffering is needed at the client to compensate for the time it takes to handoff. Resource management framework we developed takes care of scheduling so that handoff is seamless from the application level, and at the same time power is conserved.

Table 3: Handoff from BlueTooth to 802.11b

1.	BlueTooth link breakage detection	700 ms
2.	802.11b link and monitoring setup	98 ms
3.	DHCP to configure 802.11b link	1789 ms
4.	Proxy API processing time	6 ms
5.	WPAD, DHCP to get Config. URL	109 ms
6.	WPAD, query proxy.pac via HTTP	10 ms
7.	Parsing, connect to upstream proxy	4 ms
	Total Elapsed Time:	2716 ms

Lastly, we performed measurements with SeS and IPAQ/Linux client that support both 802.11b and Bluetooth interfaces. The power measurements are collected with a DAQ card at 10ksamples/sec. We have used TCP for all data communications and bnep for Bluetooth. Results for an MPEG4 video (320x160 clip) running at 15frames/s, shown in Figure 7, highlight savings of 65% in energy consumption when using our resource management (RM) over the best possible savings with 802.11b MAC layer power management (802.11b PM), with no degradation in QoS. MAC layer power management for 802.11b typically achieves significantly smaller savings due to broadcast traffic. Measurements presented in [8] show that in medium to heavy broadcast traffic, 802.11b PM achieves at most 10% savings in energy consumption. In contrast, our resource management algorithm does not suffer from the broadcast traffic problem. As a result, its savings can be significantly higher in realistic conditions than in ideal conditions shown in Figure 7 for 802.11b PM.

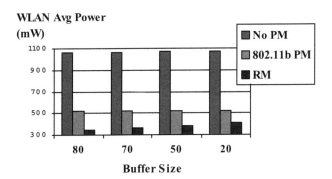

Figure 7: WNIC power consumption for MPEG4 video

In another experiment, we analyzed the performance of RM when the application data consumption rate changes by creating an application trace consisting of MP3 audio, email, telnet, WWW and MPEG2 video. We found that RM offers a factor of 2.9 times improvement in power savings over just employing Bluetooth with park mode, and a factor of 3.2 times higher than standard 802.11b power manager. Moreover, RM enhances the QoS since wireless interfaces are switched to match the data usage pattern of the application. We have also demonstrated seamless migration of connectivity from low bandwidth GPRS to 802.11 with a Linux client receiving an MP3 audio stream with similar savings while keeping MP3 decode real time.

As can be seen from the above discussion, our Smart Edge Server provides secure wireless access to the clients, has sophisticated media handling and storage capabilities and uses advanced techniques to manage resources available to it, such as bandwidth, power and the type of connectivity. As a result, the SeS platform provides immense opportunity for research and development in the area of mobility and media delivery.

7. REFERENCES

[1] Business Week, Special Issue on Wi-Fi, http://www.businessweek.com/magazine/toc/03_17/B3830003 17wifi.htm

[2] Wireless ISPs, http://www.bbwexchange.com/wisps/florida-wisps.asp

[3] Internet Content Adaptation Protocol, http://www.ietf.org/rfc/rfc3507.txt

[4] A Python ICAP Framework, http://icap-server.sourceforge.net

[5] Devaraj Das. IPSec-based Delegation Protocol and its Application, Future Trends in Distributed Computing Systems, IEEE Computer Society, May 26-28, 2004. Suzhou, China.

[6] Prakash Reddy, Venky Krishnan, Kan Zhang, Devaraj Das: Authentication and Authorization of Mobile Clients in Public Data Networks. Infrastructure Security International Conference, InfraSec 2002, Bristol, UK, October 1-3, 2002, LNCS 2437 and HPL-2002-213

[7] D. Das, G. Manjunath, V. Krishnan, P. Reddy : "Hotspot! – a service delivery environment for nomadic users system architecture", Hewlett Packard Technical Report, HPL-2002-134, 2002 and NCC2004, Indian Institute of Science, Bangalore,

[8] T. Rosing, A. Acquaviva, V. Deolalikar, S. Roy: " Server-driven Power Management", PATMOS 2003.

[9] W. Quadeer, T. Simunic, J. Ankcorn, V. Krishnan, G. De Micheli, "Heterogeneous wireless network management", PACS 2003.

[10] T. Simunic, L. Benini, P. Glynn, G. De Micheli: "Event-Driven Power Management", IEEE Transactions on CAD, pp.840-857, July 2001.

[11] Jean Tourrilhes & Casey Carter. P-Handoff : A framework for fine grained ad-hoc vertical handoff. Proc. of PIMRC 2002.

[12] Casey Carter, Robin Kravets & Jean Tourrilhes. Contact Networking: A Localised Mobility System. Proc. of MobiSys 2003.

[13] Geetha Manjunath, Venkatesh Krishnan, *A Content Adaptation Framework that supports mobility*. HPLabs Technical Report, to appear in 2004.

[14] Geetha Manjunath, Venkatesh Krishnan, *Distributed Personal Media* , HPLabs Technical Report, to appear in 2004.

[15] Devaraj Das, Geetha Manjunath, Venkatesh Krishnan, *Dynamic Web Services using Coolbase Appliance server*, HPLabs Technical Report, 2004.

[16] D. Forsberg et al, "Protocol for Carrying Authentication for Network Access (PANA)", IETF Draft, Feb 9, 2004

[17] Multi Photo Video standard, www.osta.org/mpv

[18] "How much information 2003", UC Berkeley report, http://www.sims.berkeley.edu/research/projects/how-much-info-2003/, 2003.

[19] K. Wong, H. Wei, A. Dutta, K. Young, "Performance of IP Micro-Mobility Management Schemes using Host Based Routing," WPMC , 2001.

[20] Z. Jiang, K. Leung, B. Kim, P. Henry, "Proxy Servers Based Seamless Mobility Management," WCNC, 2002.

[21] J. Tourrilhes. "L7-mobility : A framework for handling mobility at the application level," Proc. of PIMRC, 2004.

[22] D. Das, "Design and Implementation of an Authentication and Authorization Framework for a Nomadic Service Delivery System", MS Thesis, Indian Institute of Science, March 2003.

[23] M. Burrows, M. Abadi, and R. Needham, "A Logic of Authentication," Technical Report 39, DES SRC, 1990.

[24] R. Atkinson. "Security Architecture for the Internet Protocol", RFC 2401, IETF, August 1995.

Fast Authentication Methods for Handovers between IEEE 802.11 Wireless LANs

M.S. Bargh, R.J. Hulsebosch, E.H. Eertink

Telematica Instituut
P.O. Box 589,
7500AN Enschede, the Netherlands
Telephone number: +31 53 4850 485

{bargh, hulsebosch, eertink}@telin.nl

A. Prasad, H. Wang, P. Schoo

DoCoMo Euro Labs GmbH
Landsberger Strasse 308-312
80687 Munich, Germany
Telephone number: +49 89 56824 000

{prasad, wang, schoo}@docomolab-euro.com

ABSTRACT

Improving authentication delay is a key issue for achieving seamless handovers across networks and domains. This paper presents an overview of fast authentication methods when roaming within or across IEEE 802.11 Wireless-LANs. Besides this overview, the paper analyses the applicability of IEEE 802.11f and Seamoby solutions to enable fast authentication for inter-domain handovers. The paper proposes a number of possible changes to these solutions (typically in terms of network architectures and/or required trust relationships) for inter-domain operation. In addition, the paper identifies the crucial research issues therein. Possible solutions and directions for future research include: update to security infrastructure, inter-layer communication and discovery of appropriate networks.

Categories and Subject Descriptors

C.2.1 [**Computer-Comm. Networks**]: Network Architecture and Design – *wireless communication*; C.2.6 [**Computer-Comm. Networks**]: Internetworking – *Standards*.

General Terms

Design, Security, Standardization.

Keywords

Handover, inter/intra-domain, WLAN, seamless, authentication.

1. INTRODUCTION

Recently IEEE 802.11 networks, so-called Wi-Fi™ or Wireless LANs (WLANs), have gained a lot of popularity, due to its low cost and relatively high bandwidth capabilities. These networks typically offer access to the Internet and/or corporate networks for delivery of application level services such as multimedia. IEEE 802.11 networks consist of access points that can be interconnected to form a single so-called hotspot. In principle, such a hotspot has only limited geographical coverage.

In this paper, we study the problem of having continuous access to multimedia services while moving from one hotspot to another. The motivation for this study is the observation that re-establishing connectivity to a new access point, i.e., handover, takes considerable time (up to 1 second) in WLANs. This is not acceptable for a number of applications such as VoIP. Therefore, recently a number of efforts have resulted in improvements in handover delays. In particular, IEEE has finalized the Inter-Access Point Protocol (IAPP) for inter-access point communication. The IAPP, as standardized in IEEE Std 802.11f [17], enables (fast) link layer re-association at a new access point in the same hotspot. On the network layer, similar efforts are ongoing on in the IETF Seamoby group [18][4], which defines different protocols to reduce network discovery and reconfiguration delays.

As its first core contribution, this paper gives in Section 3 a comprehensive overview of the IEEE 802.11 handover process and in Section 4 an extended overview of fast authentication methods for WLANs. Since authentication contributes significantly to the total handover delay, the paper further studies the issue of accelerating the authentication process when roaming across hotspots and administrative domains. As its second core contribution, the paper addresses how to extend the fast authentication methods to enable seamless inter-domain handovers between IEEE 802.11 based WLANs. The key aspects that need to be covered in this extension have to do with network discovery and service-level agreements between network operators where authentication, access control, and accounting options are arranged for. These are described in Section 6, for which Section 5 gives an introduction by describing issues and solutions involved in intra-domain handovers. In Section 7 we draw some conclusions and give directions for further research.

2. MOTIVATIONS AND OBJECTIVES

2.1 Problem Context

According to IEEE 802.11 standard, a mobile station, referred to as Mobile Node (MN) throughout this paper, must first associate with an access point (AP) in order to be able to communicate with other nodes via the hotspot comprising the AP. A hotspot is referred to as Extended Service Set (ESS) in the IEEE 802.11 standard terminology [9]. The association mentioned typically involves some form of authentication. The recent IEEE 802.11i standard [6] defines this authentication process, which will be described in more detail later. In one ESS, where multiple APs are present, the IEEE 802.11 standard defines the re-association procedure that enables reusing part of the authentication-state from the original AP. This reuse, however, is not so evident for authenticated re-association at an AP in another WLAN.

End-to-end connectivity is provided via the IP-layer. Therefore, the ESS is part of an IP subnet in which at least one router is present. A number of IP subnets constitute an access network or a domain that is administrated by a network operator. An example of two access networks that each consists of several IEEE 802.11 ESSs is depicted in Figure 1. The IP architecture of a typical access network comprises at least one router: the Access Router (AR). Often also a gateway router (Access Network Gateway; ANG) is distinguished that separates an access network from other IP networks [11]. Connected to one or more IP subnets/ESSs, an AR is an IP router in the access network that may include access network specific functionalities, for example, related to mobility and/or QoS. In a small network, of course, the AR may take the role of gateway as well.

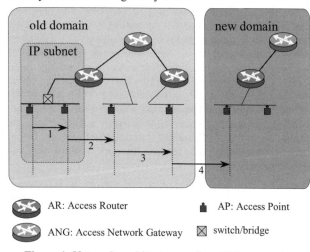

Figure 1: Network architecture and mobility scenarios.

Figure 1 also illustrates 4 categories of handover scenarios that will be used in the remainder of this paper. In scenario 1, the MN changes APs connected to the same AR's interface (intra-ESS mobility). This handover is transparent to the routing at the IP layer (L3) and appears simply as link layer (L2) mobility. In scenario 2, the MN mobility changes the AR's network interface to the MN. That is, the serving AR remains the same but routing changes internal to the AR. In principle, the MN can even keep its IP-address. In Scenario 3, the MN changes ARs inside the same domain. This requires higher hierarchical routers to deal with the locally changed IP-address of the MN. In Scenario 4, the MN moves to a new network domain (inter-domain mobility). This handover involves the assignment of a completely new IP address to the MN and invocation of L3 mobility management procedures.

For each handover scenario two aspects are of importance: the latency involved during the handover process and the security of the handover. Moreover, both aspects are closely related to each other as security mechanisms have a high impact on the overall handover latency. In the sequel, we will analyze the handover process in more detail. First, we will have a look at the involved trust relationships needed for secure WLAN handovers.

2.2 Trust Model

To design a security solution for intra- and inter-domain handover or evaluate its performance, the trust relations between the entities involved must be identified. In a public WLAN access scenario, we have one or more access providers, each operating the networks in their own domain, and several MNs, each subscribed to at least one network operator. The MNs typically do not trust each other. Also, an access provider cannot trust any MN that tries to connect to the network. The access providers may choose to trust each other. Such trust relations normally rely on (legally binding) roaming agreements or other partnerships such as a Single Sign-On federation. In such a roaming model, however, MNs may connect to the networks owned by providers that are unknown to the MNs, i.e., a direct trust relationship may not exist between these roaming MNs and providers. Therefore, an appropriate trust infrastructure is required to establish these relationships indirectly on the basis of roaming agreements.

All these trust requirements ask, among others, for security mechanisms to mutually authenticate the network and the MN at access time before granting network access [3]. This can be based on either public/secret key cryptography or other Security Associations (SAs). A SA is a relationship between two or more entities that describes how the entities will utilize security services to communicate securely. In WLANs, there are three entities involved in the authentication process: the MN, the AP and the Authentication Server (AS). The AS is an entity that resides in the access network that may participate in the authentication of MNs and APs in the WLAN. The IEEE 802.11i task group, that is concerned with IEEE 802.11 security enhancements (see also Subsection 3.1), has made the following assumptions with respect to trust during mobility: (a) a MN trusts the (home) AS and the AP with which the MN is associated, whereas (b) the MN does not trust all non-associated APs in the first instance [10].

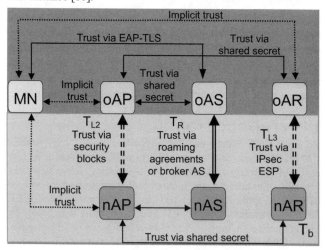

Figure 2: Trust relationships before and during a handover.

Figure 2 depicts a trust model with the trust relationships between different entities when a MN associates with an AP, denoted as old AP (oAP), and then roams to new AP (nAP). The other entities of the trust model that may play a role are old AR, new AR old AS and new AS (designated by oAR, nAR, oAS and nAS, respectively). In intra-domain handover, i.e., mobility scenarios 1, 2 and 3, the oAS and nAS entities are the same. In intra-domain handover scenarios of 1 and 2, moreover, oAR and nAR entities are the same. In Figure 2, the solid lines represent explicit mutual trust relationships while the dotted lines trust represent implicit relationships. Implicit relationships must be created dynamically to gain access to the network media. To create the implicit trust relationships and dependent of the mobility scenario, one may make use of the trust relationships between oAP and nAP, between oAR and nAR, and between oAS and nAS (designated by T_{L2}, T_{L3} and T_R, respectively, in Figure 2). In the following subsections, we sketch how these implicit trust relationships are established in two modes of a MN's operation: before (right after when the MN gets powered on) and during a handover.

2.2.1 Before a Handover

When a MN gets powered on within a domain, the implicit L2 trust relationship between the AP and the MN is transitive and derived from the fact that (a) the MN and the AS mutually trust each other (possibly via the home AS of the MN) and (b) the AS and AP trust each other (usually via a shared, preconfigured secret). The MN and the AP, on the other hand, cannot trust each other until they have securely determined that each party knows a common secret. Since the MN and AP both trust the AS, these relationships will be used to establish a trusted relationship between the MN and AP. In order to establish that trust relationship, the AP and MN must convince each other that they are who they claim to be, and must mutually derive the necessary encryption and authentication keys based upon keying material from the mutually trusted AS. For this, IEEE Std 802.11i is devised, which is explained in the next section.

An implicit L3 trust relationship must be established between the MN and the AR. Here we have a similar situation as in establishing the L2 trust relation, where two entities share a trust relationship with a third entity: the AP generally trusts the AR and the MN trusts the AP (on the basis of the L2 implicit MN-AP trust), therefore, the MN may trust the AR.

2.2.2 During a Handover

During a handover, i.e., when a MN roams from the cell area of oAP to nAP in Figure 2, new implicit L2 and L3 trust relationships must be established between the MN and nAP and the MN and nAR. The new implicit trust relationships could be created from scratch, the same way as outlined in the previous subsection. Such an approach, however, would not be efficient because the existing implicit and explicit trust relationships are not fully exploited. These trust relationships particularly include those established before the handover and those between peer entities. The trust relationships of the latter type are between oAP and nAP, between oAR and nAR, and between oAS and nAS: (a) When both oAP and nAP belong to the same hotspot, the IEEE Std 802.11f [17] foresees a means of creating a secure communication channel T_{L2} between them. (b) During an intra-domain handover and when the AR changes, the Seamoby WG mechanisms [18][4] exploit the trusted channel T_{L3} between oAR and nAR. (c) During an inter-domain handover when the AS changes, a roaming agreement T_R between two domains establishes a trust relationship between oAS and nAS.

Note that cases (a) and (b) only cover intra-domain handovers, i.e., IEEE 802.11f and Seamoby are not directly applicable for the inter-domain scenario. Moreover, note that IEEE 802.11f only cover intra-ESS handovers. In Section 4 we describe how different authentication methods establish the implicit L2 trust

relationship between MN and nAP (mainly for the authentication purposes). Some of the methods use T_{L2} and T_R to reach this objective, if the mobility scope is within an ESS and between two domains, respectively, as described above.

Establishing an implicit L3 trust relationship between the MN and nAR may follow that of the L2 on a similar way as described in the previous subsection, i.e., the nAP trusts the nAR and the MN trusts the nAP. Alternatively, an implicit L3 trust relationship between MN and nAR can be created transitively from the previous trust relationship between MN-oAR and the T_{L3} between oAR and nAR, if applicable. Note that the trust relationship of T_{L3} is not so obvious even if both oAR and nAR belong to the same administrative domain. This trust relationship must be established by using SAs to guarantee secure reverse address translation (which is a new responsibility for the nAR) and context transfer (that includes all relevant information) [4]. To actually improve the performance for facilitating secure mobility at L3 the following trust assumptions are made: (a) The nAR and oAR share a strong trust relationship. When both ARs are located in the same administrative domain, there is usually such a strong trust relationship. (b) The nAR and oAR share a secret, usually exchanged by the Internet Key Exchange (IKE) protocol. A secure channel exists between the nAR and oAR. IPsec must be provided, using ESP or AH header support with non-null encryption. In case these assumptions are invalid the whole environment is considered to be insecure

2.3 Scope and Contributions

One of our objectives in this paper is to investigate the issue of having a T_{L2} between the APs of different ESSs when transitively establishing the implicit MN-nAP trust. This is the case when an intra-domain handover (scenarios 2 and 3) or inter-domain handover (scenario 4) occurs, for which we will first sketch a number of solutions to create such an inter ESS T_{L2} relationships. We will propose how to possibly establish such a T_{L2} trust relationship, assuming that (a) a T_R exists when an inter-domain handover occurs i.e., when the AS changes, (b) a T_{L3} exists when the AR changes and an intra-domain handover takes place (or a T_{L3} can be established when an inter-domain handover occurs. To this end, IEEE 802.11f and Seamoby solutions will be explored.

Subsequently, we will investigate the usefulness and limitations of such a T_{L2} trust relationship in case of inter-domain handovers. This investigation leads to a number of guidelines indicating the necessity of complementary measures or alternative approaches. Specifically, network discovery and dynamic establishment of roaming agreements (proactively) emerge as major requirements in enabling fast inter-domain handovers. Before going into fast authentication solutions, we first give an overview of the actual handover process itself in the next section.

3. OVERVIEW OF HANDOVER PROCESS

MN movement triggers a L2 or L3 handover to take place. This section provides an overview of the L2 and L3 handover processes. For the L2 handover process, the description here is based on the IEEE 802.11 standard and its security protocol suite of IEEE 802.11i. For the L3 handover we will briefly describe the general steps needed without having a specific protocol in mind.

3.1 L2 Handover in IEEE 802.11 WLANs

When a MN moves out of the coverage area of a oAP, the MN re-associates with a neighbour nAP, where both APs belong to the same ESS. The mechanism or sequence of messages between a MN and the APs that results in a transfer of physical layer connectivity and state information from oAP to nAP is referred to as a L2 handover. During a L2 handover, the MN is mostly unable to exchange the data traffic via its oAP and nAP. This interval of no communication is often referred to as handover delay or latency. The MN initiates the so-called re-association service to carry out the IEEE 802.11 handover process. The handover process can be split into three phases: detection, search and execution, as illustrated in Figure 3 according to [8] and [21].

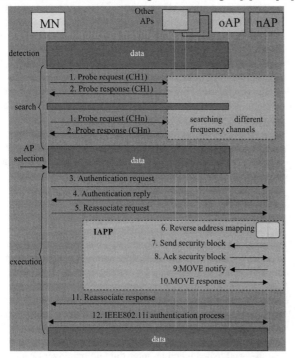

Figure 3: Handover process based on IEEE 802.11/802.11f.

3.1.1 Detection Phase

In this phase the MN detects the need for a handover, which generally depends on system design, network deployment and the requirement set by network operator or user. The need for handover can be based on: (1) detecting lack of connectivity via the current AP (in other words, detecting the loss of expected traffic) or (2) detecting a more suitable (e.g., cheaper) connection.

3.1.2 Search Phase

The search phase includes a set of so-called scans performed by the MN to find the APs in range. The standard mandates scanning of all IEEE 802.11 channels. Some implementations, however, allow scanning to be carried out for a subset of channels (in this case the MN should somehow be provided with the list of channels for scanning). On each channel, the IEEE 802.11 standard specifies two scanning modes: active and passive. In passive scanning, MNs listen to each channel for the beacon frames announcing the services of a particular AP. This may induce long delays (more than 1 second). In active scanning, MNs broadcast a probe-request management frame in each channel and wait for probe responses generated by APs, as illustrated by

messages 1 and 2 in Figure 3. Active scanning is usually faster, but consumes more power and bandwidth.

3.1.3 Execution Phase

The handover execution phase is a two-step process: authentication and re-association. Typically the nAP needs to authenticate the MN before the re-association succeeds. The IEEE Std 802.11 [9] specified two authentication algorithms: open system and shared key. These have been superseded by IEEE 802.11i that deals with security flaws of the open system and shared key solutions. The messages of IEEE 802.11i are exchanged after the re-association process and therefore we will describe them at the end of this section. To maintain backward compatibility, however, IEEE 802.11i allows open system authentication [6], as indicated by messages 3 and 4 in Figure 3. Then, the MN sends a reassociate request to the nAP (see message 5 in Figure 3). The reassociate request management frame includes, among others, the MAC address of the oAP (also called the Basic Service Set IDentifier, abbreviated as BSSID), the ESS identifier (also called Service Set IDentifier, abbreviated as SSID), and the MAC address of the MN.

The reassociate request message at the nAP triggers the IAPP-MOVE protocol sequence between the source and target APs of oAP and nAP, respectively. The MOVE protocol sequence [17] verifies the MN's association at the oAP, transfers the MN context information from oAP to the nAP (thus allowing, for example, faster re-authentication of a MN on re-association) and updates the forwarding table at L2 devices (thus, allowing frames destined for the MN to be delivered via the nAP). As the first action of the MOVE protocol sequence, the nAP finds out the IP address of the oAP based on the oAP MAC address (indicated by action 6 in Figure 3). Before oAP and nAP engaging in communication and the context transfer mentioned, IAPP requires a trusted connection between nAP and oAP. This can (dynamically) be created using Security Block (SB) items that contain information for securing the oAP-nAP connection (messages 7 and 8 in Figure 3). Now both APs can encrypt all further packets exchanged between the APs.

Having the IP address of the oAP, the nAP sends an IAPP MOVE-notify message via the secure connection to the oAP, which includes the MN MAC address (messages 9 in Figure 3). The oAP checks its association table over the MN's association. If the association exists, then the oAP will forward the MN's any relevant context to the nAP (message 10 in Figure 3); and (internally) issue a disassociation for the MN. The nAP sends a reassociate response to the MN indicating the result of the requested re-association (successful or unsuccessful). In addition, the nAP broadcasts a L2 Update Frame to update the forwarding table of L2 devices with the source MAC address of the MN.

A malicious MN should not be able to re-associate (hijack session) or disassociate (perform DoS attack) on behalf of the associated MN. Thus, the authenticity of the re-association or disassociation requests must be guaranteed. As the MN drives the re-association process, it is important that the MN proves its relationship with the oAP to the nAP. Ciphers such as TKIP or CCMP can be used to authenticate, integrity and replay protect and encrypt the re-association/disassociation frames. Alternatively, a so-called Authenticator Information Element (IE) can be added to the frames (see [10] for more information).

Because wireless networks are by definition vulnerable to snooping and possible intrusion by 3rd parties, mutual authentication and trust needs to be established (and re-established when roaming) between APs and the MN. For that purpose, IEEE 802.11i prescribes the use of EAP/IEEE 802.1X mechanisms [6]. When operating IEEE 802.1X, a Master Key (MK, also called AAA-key in the IEE 802.11i draft) and a Pairwise MK (PMK) are generated via the exchange between the MN and AS. The MK is obtained from the MN-AS authentication during the EAP-exchange. The PMK may be generated from the MK (by using the first 32 octets of the MK [7] [6]), or it may be generated by some other means (e.g. randomly chosen by the AS) and protected in transport from the AS to the MN using material derived from the MK. The actual procedure for the generation of these keys depends on the EAP-method chosen. The simplest method is EAP-MD5 challenge, which only authenticates the MN to the AS. More advanced methods, e.g., EAP-TLS, EAP-TTLS, EAP-SIM (protocol to enable WLAN authentication using any GSM phone "SIM" smart card), provide for mutual authentication of the MN and AS. Note that these methods will not authenticate the AP to the MN. When the PMK is generated, the AS pushes it securely to the AP via RADIUS transport. Delivery of the PMK to the AP is delivering the decision that the IEEE 802.1X authentication process was successful and is an indication that the IEEE 802.1X/EAP process is completed.

Subsequently, the AP and the MN use the PMK to derive, bind and verify a fresh Pairwise Transient Key (PTK). This is done via the so-called 4-way handshake. The 4-way handshake proves the liveness of both peers, demonstrates that there is no man-in-the-middle, and synchronizes pair-wise key use. The AP uses portions of the PTK to securely distribute a Group Transient Key (GTK) to all associated APs. A 2-way handshake is used for this purpose. The GTK is used to secure broadcast messages to the APs. Either TKIP or CCMP can be used to provide for data protection. All IEEE 802.11i operational phases are shown in Figure 4.

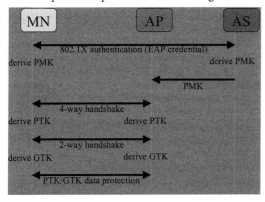

Figure 4: IEEE 802.11i operational phases.

3.2 L3 Handover

If an IP subnet change occurs during a handover from oAP to nAP, then an IP level, i.e., L3, handover will take place. This may result in a change of AR from oAR to nAR. The actions carried out during an IP level handover can be grouped into three phases of movement detection, MN configuration and path redirection, described here shortly.

In L3 handovers, a change of IP subnet is detected in the movement detection phase. When the MN establishes (or wants to establish) L2 connectivity to a nAP, the actions of this phase determine whether the nAP belongs to a new IP subnet. When the nAP belongs to a new IP subnet, the MN must obtain subnet specific parameters such as the new IP address, referred to as Care of Address (CoA) throughout this paper, in the configuration phase. This configuration enables the MN to communicate via the new IP subnet. Finally, the path redirection phase includes measures to redirect the data flow to the acquired new CoA by for example informing the correspondent nodes of the MN over the new CoA. The actions in these phases incur an additional delay on top of the L2 handover delay that should also be considered for seamless handovers across domains and WLANs. Joint optimization of L2 and L3 handover processes is perhaps the direction for achieving this objective.

4. FAST AUTHENTICATION METHODS

Research indicates that a substantial amount of latency is incurred during the exchange of authentication messages. The negative impact of in particular IEEE 801.1X authentication on the performance of the handover process has been recognized and a number of enhancements have been proposed. Relevant L2 solutions to reduce authentication latency are described in this section for intra-ESS handovers.

4.1 Proactive Key Distribution

Pro-active key distribution intends to reduce the latency of authentication phase by pre-distributing key materials one hop ahead of a MN [20] for EAP-TLS authentication. For each access, a Neighbour Graph captures dynamically the set of neighbour APs. Assume a MN is associated with the oAP and as a result of the MN-AS authentication; an old PMK is shared between the MN and oAP. The following message sequence is prescribed for pro-active key distribution: (1) The AS constructs the new PMK from the old PMK, MAC address of the MN and of the nAP, and sends it to the nAP (this step is repeated for all APs in the Neighbour Graph of the oAP, which is known at the AS); (2) When the MN roams to the nAP; the MN derives the new PMK the same way as the AS did; (3) MN and nAP derive a new PTK via key agreement, e.g., via the 4-way handshake and a new GTK.

If either the nAP or MN does not support pro-active key distribution, a full re-authentication must be done. The AS of the ESS initiates pro-active key distribution. The authentication latency, attributed to the process described in the previous section, is reduced from an average of 1.1 sec to 25 msec with pro-active key distribution and Neighbour Graph [20]. It also eliminates problems with sharing key material amongst multiple APs. Disadvantages of this method are its heavy burden on the AS server and its requirement for new RADIUS messages. We look at the applicability of proactive key distribution in inter-domain handover in Section 6. The trust model of pro-active key distribution is transitively constructed from MN-AS, MN-oAP (as exists in old PMK) and AS-candidate nAPs trust relationships.

4.2 Pre-authentication

IEEE 802.11i [6] defines pre-authentication for roaming users. Pre-authentication can be useful as a performance enhancement, as the L2 execution phase will not include the protocol overhead of a full re-authentication when it is used. Pre-authentication relies on IEEE 802.1X. A MN can initiate pre-authentication whenever it has an association established with an oAP. To effect pre-authentication, the MN sends an IEEE 802.1X EAPOL-Start message as a data frame to the BSSID of a targeted nAP (within its present ESS) via the oAP with which it is associated currently. The targeted nAP then may initiate IEEE 802.1X authentication to the MN. The distributed system will be configured to forward this message to the AP with which the MN is associated. The pre-authentication exchange ends when, after deriving the new PMK, the nAP sends the first message of the 4-way handshake. When pre-authentication is used, the MN and nAP must cache the new PMK that has been derived during the IEEE 802.1X pre-session. An identifier must be included in the reassociate-request/response messages to exchange information whether or not both MN and nAP support pre-authentication, i.e. have cached new PMKs.

The benefits of pre-authentication are that the authentication phase becomes independent of roaming and that the MN may authenticate to multiple candidate nAPs at the same time. On the negative side, pre-authentication introduces new opportunities for DoS attacks, requires fat APs and MNs, and might unnecessarily burden the AS server (unless new PMK caching is used). Pre-authentication cannot occur beyond the first AR due to the fact that EAPOL packets are used to carry authentication information. This implies that pre-authentication is not suitable for roaming scenarios 3 and 4. The trust model of the pre-authentication method is transitively constructed from MN-AS and AS-candidate nAPs trust relationships.

4.3 Predictive Authentication

Pack [15] has proposed the predictive authentication protocol by using a modified IEEE 802.1X key distribution model. Instead of the one to one MN-AS message delivery of IEEE 802.1X, the model of [15] supports one to many authentications. The MN sends an authentication request to the AS, and the AS sends multiple authentication responses to all APs within a certain Frequent Handoff Region (FHR). This FHR is a set of adjacent APs and is determined by the AP's locations and user's movement patterns. The trust model is similar to that of pro-active key distribution and has the same benefits and disadvantages. The introduction of the FHR theoretically allows for inter-domain roaming, in practice, however, the implementation of a FHR over multiple domains may cost a lot of effort.

4.4 L2 Context Transfer Using IAPP

An important aspect of a handover that has recently drawn the attention of research community is the transfer of context information. IAPP MOVE operation enables transferring a MN's context information from the oAP to nAP securely. In practice, the IAPP MOVE protocol sequence uses IPsec to establish a secure channel between the nAP and oAP, i.e., the T_{L2} trust relationship in Figure 2. Here we describe how to use IAPP context transfer features for fast authentication in reactive and proactive operation modes.

4.4.1 Reactive Mode

The context exchanged in IAPP MOVE operation may include security credentials whereby the implicit trust relationship between MN and nAP can be built on top of the implicit MN-oAP trust relationship and T_{L2}. When a trust relationship between the MN and oAP was established, both MN and oAP share a common secret (e.g., the old PMK). The context transfer of IAPP should not enable nAPs to obtain the keys used for encrypting traffic on the oAP, because otherwise a rogue nAP can decrypt traffic previously captured on the oAP. Several mechanisms are

available to prevent this, e.g., the oAP transfers to the nAP a transfer key derived via a one-way function from the old key or, alternatively, the oAp derives a new key that does not depend on the old key.

The latter mentioned example solutions, however, allow the oAP to be aware of the common secret between the MN and nAP (after all it will be based on the key transferred in the MOVE response). This may conflict with the IEEE 802.11i requirement that all APs other than the one with which the MN is associated are not trusted [6]. However the trusted oAP can usually observe the L2 traffic on the common Distribution System (DS) including those exchanged between the MN and nAP. Thus, no additional risk is introduced by these solutions.

4.4.2 Proactive Caching

For interactive, or high quality, real-time applications the performance of the context transfer within the IAPP MOVE protocol sequence is not sufficient. Thus IAPP provides the IAPP CACHE protocol sequence to actually move the state to a set of nAPs, i.e. neighbouring APs, before the MN tries to re-associate at one of those nAPs. The set of neighbouring APs relative to an AP is called the Neighbour Graph whose implementation is vendor dependent according to the standard.

Proactive caching can eliminate the time needed to transfer the context information by the MOVE protocol sequence (after a re-associate request). Ref. [21] reports an order of magnitude improvement (from 15.37 msec to 1.69 msec) for performing the re-associate process of Figure 3 when proactive caching is used. When a MN re-associates with a nAP, the nAP looks up its cache for the context entry of the MN. If the context exists and is not expired (a cache hit), then the nAP sends a re-associates response message to the MN. Otherwise (a cache miss), the nAP invokes the IAPP MOVE protocol sequence. In any case, a new IAPP CACHE protocol sequence is invoked to push the latest context to the neighbours of the nAP.

5. EXTENSION OF INTER-ACCESS POINT TRUST RELATIONSHIP

As mentioned in Subsection 4.4, the existence of inter-AP trust, as denoted by T_{L2} in Figure 2, improves the performance of L2 handover. However, the scope of IAPP protocol realizing the T_{L2} in WLANs is limited to a single ESS and it cannot be applied to inter-ESS or inter domain handovers (i.e., scenarios 2, 3 and 4 in Figure 1). In order to use IAPP between the APs of different ESSs and/or different domains, we observe three main obstacles: (1) reverse address mapping: to map the MAC address of oAPs to their IP addresses, (2) IP address translation: to map between private IP addresses of APs and routable IP addresses, and (3) trust and the security issues between APs. The first obstacle stems from the fact that MNs obtain the MAC addresses of candidate nAPs, while the communication between APs in IAPP is mainly IP based. The second issue arises from the fact that the IP addresses of APs will likely be private in IPv4 settings. The third issue is a known requirement for any context transfer process.

This section sketches a number of scenarios to transfer L2 context between APs in inter-ESS handovers, using IAPP and Seamoby protocols. We assume here that measures are taken for arranging Security Blocks between the adjacent APs of two ESSs/domains. For inter-domain handover, this can be arranged during establishing a roaming agreement, as denoted by T_R in Figure 2.

5.1 Direct L2 Context Transfer

This subsection discusses a simple scheme to directly transfer L2 context between APs in inter-ESS handovers. With a re-associate request message the MN presents the MAC address of the oAP to the nAP. To activate the context transfer feature of IAPP, the nAP should know the IP address of the oAP. This reverse address-mapping problem can be resolved by, for example, using a roaming server. This is similar to the RADIUS server as proposed for intra ESS handovers by IEEE Std 802.11f. The emergence of the address translation issue to transfer L2 context between APs of different ESSs depends on the network architecture. We consider two scenarios: APs have private or public addresses.

In cases where the range of available public IP addresses (i.e., directly routable and reachable within the Internet) is limited, as in the case of IPv4, the APs are likely assigned private IP addresses. The private IP addresses are hidden to the Internet and therefore a Network Address Translation (NAT) is used to convert private IP addresses to public IP addresses and vice versa. The most general case is the one in which APs are behind different NATs, and therefore NAT traversal mechanisms are necessary to establish an IAPP connection between the AP. To carry IAPP MOVE protocol sequences between nAP and oAP of different ESSs two NAT operations are needed for converting between private and public IP addresses of the APs. This enables MOVE packets to traverse the network segments between two ESSs.

With advent of IPv6 the range of available IP addresses will enormously increase and therefore it is feasible to provide every AP with a public IP address. There is also an alternative IEEE 802.11 deployment scenario where each AP is integrated with exactly one AR, see for example [16]. In this case, the AP/AR has a public IP address. When both nAP and oAP poses public IP addresses the issue of address translation vanishes by definition.

5.2 Encapsulating L2 Context in L3 Context

This section provides a scheme for transferring L2 context between APs in inter-ESS handovers that involves integration of the Context Transfer Protocol (CTP) [18] and the IAPP. The scheme embeds a L2 IAPP context in a L3 context and transfers the result using the CTP. The CTP is being defined by the Seamoby Working Group of IETF to transfer context information, i.e., referred to as L3 context in this paper, from a oAR to a nAR. This scheme resolves the address translation issue by exploiting the IP addresses of ARs, which are inherently used for L3 context transfer. The reverse address mapping issue, however, emerges in another form here where the MAC addresses of candidate nAPs are mapped to the IP addresses of the corresponding nARs.

Consider the inter-ESS mobility scenario of 3 in Figure 1. In this handover, within one domain the ESS and AR of the MN change. The following message sequence transfers L2 context within L3 context, as illustrated in Figure 5. The described scheme is based on the "network controlled non-predictive L3 context transfer" approach of [18] and a modified 802.11 re-association service. (1) MN sends a re-associate request to the nAP that includes the oAP MAC address and the MN MAC address. The nAP relays the IAPP MOVE-notify message to the nAR; (2) nAR maps the oAP MAC address into the oAR IP address, using the inverse address translation scheme as provisioned by Seamoby for the Candidate Access Router Discovery (CARD) protocol [4], and nAR sends

the move-notify message embedded in a L3 Context Transfer Request (CTR) message of the CTP to the oAR; (3) oAR checks an authorization token (issued by the MN) for L3 context transfer and delivers the MOVE-notify part of the L3 context to the oAP; (4) oAP checks whether there is a valid record of the MN's association and sends a MOVE-response message containing the L2 context of the MN to the oAR; (5) oAR embeds MOVE-response message in a Context Transfer Data (CTD) message of CTP and sends it over to the nAR which extracts the L2 MOVE-response message from the L3 context and sends it to the nAP; (6) nAP examines the status of the MOVE-response and if it is successful, it initiates an ADD-notify (see the description below) on the new ESS and sends a reassociate reply message to the MN; (7) MN goes on with L3 configurations and path redirection to the new subnet (e.g., by acquiring a new CoA and doing the Mobile IP registration) and sends a Context Transfer Activate Request (CTAR) to the nAR.

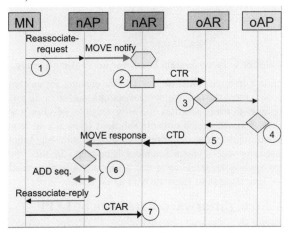

Figure 5: A scheme transferring L2 context within L3 context.

In step 1 the reassociate-request is used, opposed to the way done currently, to allow the new ESS to look for the context of the MN on the old ESS. Moreover, this reassociate-request should seem as the associate-request of the current IEEE Std 802.11. Thus, step 6 requires that an IAPP ADD notify message [17] to be also triggered upon the reception of the result of the MOVE response. The notification message informs other APs on the DS of the new ESS over the association. Alternatively, one may have used a modified associate-request message with the oAP MAC address (not dealt with here). In steps 2 and 5, the MN L3 context is identified by the MAC address of the MN, while the CTP requires that the old IP address of the MN is used. Although this does not conform to the drafted CTP standard, it uniquely identifies the MN at the oAR (because the oAR should have the MN MAC address to MN old IP address mapping), as well as at the nAR and nAP. The scheme described above is for scenario 3, where an intra-domain handover occurs. In inter-domain handovers, i.e., scenario 4, the CTP of Seamoby does not hold because the scope of Seamoby is for handovers within one administrative domain. The following section elaborates more on this issue.

6. FAST AUTHENTICATION IN INTER-DOMAIN HANDOVERS

In this section we will look at inter-domain aspects of fast authentication methods. Inter-domain means that we explicitly look at cases in which MNs move from one visited (i.e., old) domain to a new domain, where the authentication is typically proxied towards the home domain of MNs. Of course, in practice the home domain can be the same as the old or the new domain. This situation is depicted in Figure 6. Inter-domain handover is not only a technical issue. A lot of business and trust aspects play a role. In the sequel of this section, we assume that there is an intention between the two domains, i.e., a roaming agreement, to have inter-AP or inter-AR context transfer features. Note that the IAPP and Seamoby solutions are just for intra-domain operations.

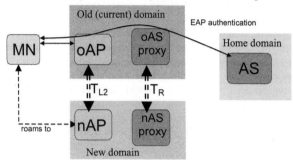

Figure 6: Inter-domain handover and authentication.

Being in a creative mode, we try to define changes for existing (or proposed) mechanisms, deployment options, and so on in order to make inter-domain seamless mobility feasible in some way or another. We will sketch possible solutions (typically in terms of network architectures and/or required trust relationships) that can be used for inter-domain seamless handovers. In this section, where applicable, we assume that a MN actually always tries a reassociate request, to associate with a new ESS.

6.1 Straightforward Extension of IAPP

A simplistic approach for extending IAPP for inter-domain mobility is to explicitly off-line connect specific APs between two domains using a secure connection (typically via a point-to-point secure connection). Hence, it requires an explicit network architecture that must be configured by the operator of the network (and may also be governed by SLAs, of course). This situation is depicted Figure 7. The approach here is to use the pre-existing secure link (either at the network or transport Layer) to exchange context information using similar mechanisms as IAPP. For L2 aspects, this means that oAP must know and communicate with nAP (and vice-versa). This is similar to the standard operation of IAPP. Making this secure connection automatically is also possible but it may involve a lot of control traffic, as it is likely that both APs cannot easily be reached from outside their domain. For example, approaches mentioned in Section 5, can be deployed. In case of encapsulating a L2 context in a L3 context, the Seamoby protocols of CTP and CARD should be extended for inter-domain handovers. This can be based on the fact that "if two domains trust each other, they allow topology information sharing and context information exchange across adjacent ARs".

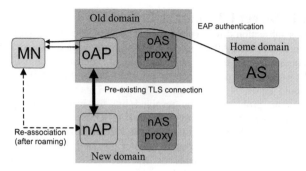

Figure 7: Off-line connecting APs.

The reassociate request message can be used to convey the information in a similar way as done for the intra-domain case, and (a derivate of) the PMK can be exchanged as part of the IAPP protocol exchange between the oAP and nAP. The main problem with this approach is that of trust and transitivity. Direct authentication between the MN and nAP is not performed in the new domain because the security context in oAP does not contain any explicit reference to the identity of the MN (that is shared only between the MN and the home AS, and governed by trust relationships between the oAP and oAS proxy, and oAS proxy and home AS). Hence, the new domain accepts any user from the old domain, based on the credentials that were presented at the original association of the MN at the old network. This leads to a complete chain of trust that only implicitly exists, and cannot be (re-)checked by the new domain at all. This is, in our view, not acceptable if the new domain does not have a roaming agreement, i.e., trust relationship T_R in Figure 2, with the home domain. For inter-domain roaming, we therefore always require that either (a) the home AS should get involved in order to control authenticated access to the network and to ensure proper accounting, or (b) the new domain should have a roaming agreement with the home domain if a transitive trust is created. The latter option requires the discovery of the existence of a roaming agreement between the new and home domains before the handover.

As an option to extend IAPP, one can encapsulate L2 context in L3 context, see Subsection 5.2, but now across two domains. The CARD protocol [4] of Seamoby can be extended in the same way as L2 pre-authentication is used: if two domains trust each other, they can share topology information, which means that also information of external ARs can be returned to the MN. This is a way of actually directing the MN to a particular friendly nAR.

6.2 Inter-domain Proactive Key Distribution

The idea here is to exploit knowledge of handover candidates in all networks and pro-actively distribute new PMKs over different domains. This differs from the previous approach in that it involves the AS of the home domain, because this server is responsible for the creation of the required new PMKs (see also Subsection 4.1 where the intra-ESS/domain variant is described). The advantage of this approach is that re-authentication can be done with the same performance as done for the intra-ESS/domain case. The architecture is depicted in Figure 8, from which it is clear that pro-active key distribution requires three things: (1) pushing the identification of the oAP to the home AS, (2) computing the set of possible nAPs close to the oAP of the MN at the home AS, and (3) signalling from the home AS to the new domain's APs.

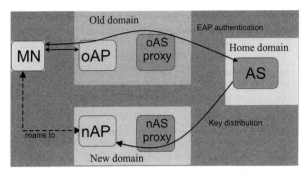

Figure 8: Inter-domain pro-active key distribution.

6.2.1 Location Information at the Home-AS

RADIUS proxies are able to remove any "non-essential" information from the RADIUS request before proxying the authentication request to the home AS. From a security viewpoint, proxies are almost always configured in such a way that the BSSID of the involved AP is removed from the request that is sent to the home AS. Of course, when this fast handover policy is to be used by the domain, this configuration can be changed to actually do transfer the BSSID to the home domain. This does not solve the complete problem; because another practical complication is that there might be several intermediate RADIUS proxies. Such a chain of proxies are used for trust-reasons: the "central" proxy maintains a collection of domains that adhere to a common access policy, and alleviates each individual domain of having to make peer-to-peer relationships and share RADIUS secrets (very cumbersome from a management point of view).

When the MN roams within a foreign domain, its home AS needs to know to which oAP it is associated, in order to predict the proper target nAPs for key-distribution. This location-update can be performed by every oAP, e.g. as a reaction on the EAPOL-start message that initiate re-association, or after the successful context transfer after an IAPP MOVE initiated at the oAP. The oAP is typically not aware of the home domain of the MN; this information can be cached at the proxy AS. Other advantages of the proxy are: the local domain has a single point of control over the information it actually transmits to the home AS, and the APs do not directly communicate with entities outside the domain.

Another, more scalable and more privacy-sensitive, option is to not regularly inform the home AS of the actual location of the MN at all, but only inform the home AS when the MN is likely to roam to another domain. This local domain knowledge can easily be made available in the local AS, and only inform the home AS whenever there is the possibility of roaming.

6.2.2 Computation of Nearest Access Points

If the home AS is aware of the BSSID of the oAP, it can simply learn distances between APs from keeping track of subsequent associations. In combination with proper caching algorithms this may lead to a practical solution, see for example [20] and [21]. Another option for reaching scalability and manageability is to involve the proxy ASs, and maintain only a mapping between oAP and oAS proxy at the home AS, and extend the RADIUS protocol to request from the oAS proxy the nearest nAPs to the oAP. Of course, a complete database can also be configured by means of management tools. If the home AS is only notified with a collection of possible nAPs and domains, it only has to check if the MN is allowed to roam to these domains, and proceed.

6.2.3 Pushing Keys Towards the New Access Point

The final step in the algorithm is to create new PMKs and push these to the nAPs. For this to happen, the home AS sends a NOTIFY-REQUEST via the nAS proxy to the nAPs. This is not supported by the current RADIUS protocol. However, there are drafts from the IRTF aaaarch group [19] that already proposed the basis for the required extensions. If the nAP accepts the request for pro-active key distribution, it can proceed using the same procedures as published in [20].

Using the pro-active key distribution method requires some changes in current RADIUS configurations/protocol and the AP behaviour. The disadvantage of this approach is that it only takes care of fast re-authentication on L2 and thus no other context element is transferred. This means that this approach solves only part of the seamless handover problem. Another disadvantage is the necessity to cache information on each AP without knowing whether it will actually be used. This leads to fat APs and ASs. Finally, the computation of fresh PMKs leads to load on the ASs.

6.3 Pre-authentication over Multiple Domains

When pre-authentication (see Subsection 4.2) is used, the initiative comes from the MN. The MN requests pre-authentication at the oAP currently associated with the MN and the oAP forwards the authentication request to the requested nAP. Effectively, EAP authentication is performed between the MN and the home AS (via the nAS proxy), and the result of the authentication is cached at the nAP. A possible extension of the protocol to the inter-domain handover case would consist of the following phases: the MN associated at the oAP pre-authenticates at nAP via the oAP; EAP authentication proceeds between MN-oAP-nAP-nAS proxy-home AS; the resulting PMK is stored at the nAP and the MN; upon re-association at the nAP, MN is authenticated automatically and PTK is derived (PMK is stored already); and finally, IAPP context transfer may optionally take place between the oAP and nAP for other types of information. This is depicted in Figure 9.

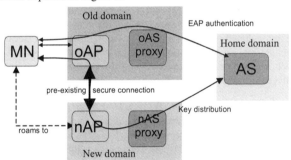

Figure 9: Pre-authentication at multiple APs.

The trust model here is similar to the IAPP model presented in Subsection 6.1: it is assumed that the two APs know and trust one another, and that they are configured in such a way that they forward IEEE 802.11i pre-authentication requests to one another across two domains (e.g., using schemes in Section 5). The difference between this approach and the pro-active key distribution mechanism is the existence of a pre-established secure link between the oAP and nAP for context transfer.

The drawback in this approach is that the configuration of the network becomes quite complex: between each couple of adjacent APs a secure connections is needed, every AP needs knowledge about nearest APs, and the MN itself needs information about the possible nAPs that it wishes to associate to. Furthermore, the authentication is governed only by roaming agreements between the old and new domains. When the new domain does not have a roaming agreement with the home domain, the pre-authentication approach may fail due to lack of trust between the nAS proxy and home AS. Finally, also the disadvantages of the L2 pre-authentication mechanisms apply.

In conclusion, it is possible to use the pre-authentication approach across domain borders. However, it does require effective mechanisms for discovery of appropriate domains and for reliable exchange of configuration information across domain borders.

7. CONCLUSIONS AND FUTURE WORK

In this paper we gave a concise overview of fast authentication solutions for IEEE 802.11 WLANs. In particular, we focused on the applicability of IAPP (IEEE Std 802.11f) and Seamoby protocols (like CTP and CARD protocol) for inter ESS mobility. The particular objective of this study was to identify possible issues when deploying these new protocols to achieve seamless inter-domain mobility. Authentication delay, network discovery and inter-layer communication were identified as the key issues of current standards and protocols for achieving seamless mobility across network domains.

Summarizing, the main results presented here are: (a) IAPP and Seamoby results cannot be applied "out of the box" for inter-domain seamless mobility; (b) Extending IAPP and Seamoby protocols for inter-domain mobility requires enhancements to the security infrastructure with mechanisms for home ASs to push authentication context information to other domains; (c) Inter-domain context transfer requires interaction with the home AS for both authentication and accounting. If any of these aspects are relevant, trust relationships between home AS and visited AS must be available or be established; (d) This implies that dynamic establishment of trust-relations between APs and between ASs is required; (e) Optimizing the communication between link and network layers may significantly improve handover delays; (f) Knowing whether the new domain has roaming agreements with the home domain of the MN in upfront requires suitable discovery mechanisms, so that, if there is such a roaming agreement in place, the MN may handover relying on the roaming agreement between the new and old domains.

8. ACKNOWLEDGMENT

The work described in this paper is part of the FISH project carried out in collaboration between DoCoMo Communications Laboratories Europe GmbH and Telematica Instituut. The authors would like to thank Hans Zandbelt, Arjan Peddemors and Remco Poortinga who reviewed an early version of this paper.

9. REFERENCES

[1] R. Stewart et.al, "Stream Control Transmission Protocol," IETF RFC 2960, Oct. 2000.

[2] J. Rosenberg et.al., "Session Initiation Protocol, baseline spec", IETF RFC3261, Jun. 2002.

[3] Y. Matsunage, A.S. Merino, T. Suzuki, R.H. Katz, "Secure Authentication System for Public WLAN Roaming", in Proc. of WMASH'03, San Diego, California, USA, Sep. 2003.

[4] M. Liebsch, A. Singh (Editors), H. Chaskar, D. Funato, and E. Shim, "Candidate Access Router Discovery", IETF draft <draft-ietf-seamoby-card-protocol-06.txt>, exp. Jun. 2004.

[5] Port-Based Network Access Control, IEEE 802.1X, 2001.

[6] Draft Amendment to … Part 11: Wireless Medium Access Control (MAC) and Physical Layer (PHY) Specifications: Medium Access Control (MAC) Security Enhancement, IEEE Std 802.11i/D10.0, work in progress, July 2003.

[7] B. Aboba, D. Simon, J. Arkko and H. Levkowetz (Ed.), "EAP Key Management framework", IETF draft <draft-ietf-eap-keying-01.txt>, work in progress, expires: Apr. 2004.

[8] J. O. Vatn, "An experimental study of IEEE 802.11b handover performance and its effect on voice traffic," Telecommunication Systems Laboratory, Department of Microelectronics and Information Technology, KTH, Royal Institute of Technology, Stockholm, Sweden, Rep. TRITA-IMIT-TSLAB R 03:01, Jul. 2003.

[9] Part 11: Wireless LAN Medium Access Control (MAC) and Physical Layer (PHY) Specifications, IEEE 802.11, 1999.

[10] B. Aboba. (2002, Jun. 17). "IEEE 802.1X Pre-authentication" [online]. Available http://www.drizzle.com/~aboba/IEEE/11-02-TBDr0-I-Pre-Authentication.doc.

[11] J. Manner AND M. Kojo, "Mobility Related Terminology", IETF draft <draft-ietf-seamoby-mobility-terminology-06.txt>, expires: Feb. 2004.

[12] O. Vatn and G.Q. Maguire Jr., "The effect of using co-located care-of addresses on macro handover latency," in Proc. of 14th Nordic Tele-traffic Seminar (NTS 14), Lyngby, Denmark, Aug. 1998.

[13] D. Johnson, C. Perkins, and J. Arkko, "Mobility Support in IPv6", IETF draft <draft-ietf-mobileip-ipv6-24.txt>, work in progress, Expires: Jun. 2003.

[14] E. Wedlund, and H. Schulzrinne, "Mobility Support Using SIP," in Proc of 2nd ACM/IEEE International Conference on Wireless and Mobile Multimedia (WoWMoM'99), Seattle, USA, Aug. 1999.

[15] S. Pack and Y. Choi, "Fast Inter-AP Handoff using Predictive-Authentication Scheme in a Public Wireless LAN," in Proc. of Networks 2002 (Joint ICN 2002 and ICWLHN 2002), Aug. 2002.

[16] P. McCann, "Mobile Ipv6 fast handovers for 802.11 networks", IETF draft <draft-ietf-mipshop-80211fh-00.txt>, work in progress, expires: Aug. 2004.

[17] IEEE Trial-Use Recommended Practice for Multi-Vendor Access Point Interoperability via an Inter-Access Point Protocol Across Distribution Systems Supporting IEEE 802.11 Operation, IEEE Std 802.11f, Jul. 2003.

[18] J. Loughney (editor), M. Nakhjiri, C. Perkins, and R. Koodli, "Context Transfer Protocol", Internet draft <draft-ietf-seamoby-ctp-08.txt>, work in progress, expires: Jul. 2004.

[19] W. Arbaugh and B. Aboba, "Experimental Handoff Extension to RADIUS," IETF draft <draft-irtf-aaaarch-handoff-01.txt>, work in progress, expires: Apr. 2003.

[20] A. Mishra, M. Shin, N.L. Petroni Jr., T.C. Clancy and W. Arbaugh, "Pro-active Key distribution using Neighbor Graphs," IEEE Wireless Comm. Magazine, Feb. 2004.

[21] A. Mishra, M.H. Shin and W. A. Arbaugh, "Context Caching using Neighbour Graphs for Fast Handoffs in a Wireless Network," in Proc of IEEE INFOCOM, Hong Kong, Mar. 2004.

Proactive Context Transfer in WLAN-based Access Networks

Ha Hoang Duong   Arek Dadej   Steven Gordon

Institute for Telecommunications Research, University of South Australia
SPRI Building, Mawson Lakes Boulevard, Mawson Lakes, SA 5095, Australia
Ha.Duong@postgrads.unisa.edu.au, {Arek.Dadej, Steven.Gordon}@unisa.edu.au

ABSTRACT

In recent years, many protocols have been developed to support mobility of wireless network nodes, e.g. Mobile IP suite of protocols designed to support IP routing to mobile nodes. However, support for truly seamless mobility requires more than just routing; every service associated with the mobile user needs to be transferred smoothly to the new access network. In this paper, we consider the problem of transferring service state (context) at both the link and IP layers. Based on the rate of SNR change in the wireless access channel, we propose a scheme that proactively transfers context information associated with the mobile user. The proposed scheme estimates the best moment in time for transferring context information, to assure the shortest waiting time of the transferred context at the new access network. We also propose and describe a new concept, forced handover, helpful in proactive transfer of context information when the mobile node moves from one access sub network to another. We present preliminary simulation results and a discussion on the performance of the proposed scheme. The scheme is shown to be helpful in ensuring seamless mobility, while the number of unnecessary handovers resulting from the proactive nature of the scheme can be kept at a controllable level.

Categories and Subject Descriptors

C.2.1 [**Computer Communication Network**]: Network Architecture and Design

General Terms: Algorithms, Performance, Design.

Keywords: Context Transfer, handover, WLAN, mobility.

1. INTRODUCTION

The common availability of third generation mobile networks (3G) and Wireless LANs have made wireless networking an increasingly important and popular way of providing Internet access to users on the move. However, the mobility of wireless

users has also created a number of technological challenges, especially when a Mobile Node (MN) changes the point of attachment to the network. In recent years, a great deal of research effort has been spent on the issue of mobility, and resulted in development of the general framework, as well as specific mechanisms and protocols supporting mobility. For example, the IETF Mobile IP Working Group has developed a solution officially named **IP mobility support** [2], and commonly known as **Mobile IP**.

Mobile IP and other mobility support protocols are intended to solve the problem of IP routing (i.e. finding the IP path) to MNs. Typically, however, the access network may also need to establish and keep service state information (service context) necessary to process and forward packets in a way that suits specific service requirements, for example Quality of Service (QoS) state or Authentication, Authorization and Accounting (AAA) state. Another example of context information may be the header compression state established and maintained between the access router (AR) and the mobile node (MN) to reduce the large IP header overhead of short (e.g. voice) packets over a bandwidth-limited wireless link. To provide truly seamless mobility for real-time applications, both the IP level connectivity and the relevant context information have to be established or re-established as quickly as possible. However, the current research [6] indicates that it is impossible to re-establish IP connectivity and service context within the time constraints imposed by real-time applications such as Voice over IP. Therefore, Context Transfer (CT) has been suggested as an alternative way of rebuilding the service context at the new access network.

The IETF Seamoby Working Group (WG) has been working on the Context Transfer Protocol (CTP) draft [7] for three years now, and intends to submit the draft as an experimental RFC in the near future. The CTP describes a simple way to transfer context information from the old AR to the new AR, so that the services can be re-established more quickly. It is expected that CTP can save time and bandwidth, and consequently improve handover performance. Although CTP is now being finalized, there are many issues yet to be resolved when employing the protocol in support of specific services. For example, as CTP only specifies the transfer procedure between two ARs, it is not clear what CTP can or should do in cases when the service involves a number of other network entities. Unfortunately, most services such as AAA, QoS, or security, require participation of not only ARs, but also other network entities. Therefore, some additional time will be required to re-establish service state after the new AR receives context information by means of CTP. This limitation can be considered a result of following a reactive CT approach, and provides a good motivation to examining the alternative proactive approach of CT.

Recently, a number of researchers have become interested in proactive IP mobility procedures. T. Pagtzis [10] suggested a proactive IP mobility model where MN's IP connectivity and other context are established at the new point of attachment in advance of the actual handover (transition between points of attachment). The key point in this model is the Mobility Neighbor Vector (MNV) – Routing Neighbor Vector (RNV) mapping. The MNV represents a collection of cells within the neighborhood reachable from the current cell; while RNV is a collection of routers associated with MNV. Discovery of the MNV-RNV mapping is achieved incrementally by means of dynamic learning i.e. MN's handover transitions between Access Points (AP) and ARs. While Pagtzis' work focused on proactive mobility at the IP layer level, Mishra in [1] focused on proactive context caching at the link layer level. In the link level model, after the MN associates with an AP, the AP will forward MN's context information to neighbor APs. Each AP learns about its neighbors through previous re-associations of mobile nodes.

The shortcoming of the above models is that the authors did not consider the waiting time of the transferred context at the new access network. Timing is an important aspect in context transfer, especially in the case of QoS context. If QoS context is transferred too early in respect to handover, resources held (reserved) in the neighboring access networks will be wasted until the MN re-establishes IP connectivity at its new point of attachment. Therefore, it is desirable that context information is transferred as close as possible to the handover time. Another key requirement for successful proactive context transfer is handover prediction. If handover prediction fails (e.g. predicted handover does not take place), network resources will be wasted.

In this paper, we propose a proactive scheme for context transfer, based on the rate of SNR change. Our scheme attempts to estimate the best time for proactive CT. We also suggest and describe a new concept of forced handover helpful in assuring that there is sufficient amount of time available for the CT to be successfully completed, so that the transferred context at the new AR is valid when the MN re-establishes IP connectivity at the new point of attachment. In this research, we examine forced handovers and proactive context transfer in WLAN-based access networks. As future work we intend to investigate these techniques in other types of access network (e.g. cellular network).

The rest of the paper is organized as follows. In the next section, we provide background information on handovers and context transfer in 802.11 Wireless LAN. Our contributions (the proactive CT scheme and the concept of forced handover) are described in section 3. Results from analysis of the scheme (via simulation) and subsequent discussions are presented in section 4. Finally, in section 0, we make some concluding remarks and comment on the areas for indented future work.

2. HANDOVERS AND CONTEXT TRANSFER IN 802.11 WIRELESS LAN

In this section, we give an overview of signal strength based handover algorithm in 802.11 WLAN, and two protocols developed by the IETF Seamoby WG, namely the Context Transfer Protocol (CTP) and Candidate Access Router Discovery (CARD) Protocol. These two protocols are expected to work closely with Mobile IP [2] to facilitate seamless handover.

Before starting the discussion, let us clarify the terminology used throughout the remaining part of the paper.

Access Point (AP) - a radio transceiver via which a MN obtains link layer connectivity to the access network.

Access Router (AR) - an IP router residing in an access network, connected to one or more APs, and offering IP connectivity to the MN.

Inter-AP handover – a process of switching (handing over) from one AP to another. In the IEEE 802.11 standard, this process is called re-association.

Inter-AR handover - a process of switching from one AR to another AR.

An inter-AP handover may result in an inter-AR handover, depending on whether the old AP and the new AP connect to the same AR or not. As an example, Figure 1 illustrates a MN roaming in an area served by AR1 and AR2. The MN encounters an inter-AP handover when it moves from the cell served by AP1 to the cell served by AP2. However, when the MN switches from AP2 to AP3, it results in an inter-AR handover from AR1 to AR2.

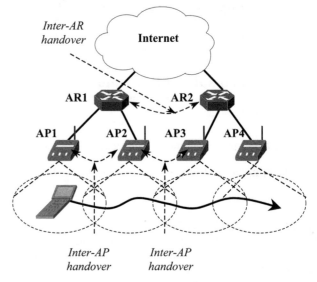

Figure 1 Inter-AP handovers & Inter-AR handovers.

2.1 Handovers between 802.11 WLAN Access Points

In the IEEE 802.11 WLAN, a MN leaving an AP is required to find the next AP and re-associate with this AP. A fundamental question is when does the MN need to switch from one AP to another. Typically, in most of implementations, for example in [8], quality of the communication link is used to make the handover decision, although in some other implementations, e.g. [3], the current load on the APs is also taken into account. In Figure 2, the typical parameter of communication quality, signal-to-noise ratio (SNR), changes between two adjacent APs, AP1 and AP2. As MN is leaving AP1 toward AP2, the SNR_1 from AP1 decreases, and the SNR_2 from AP2 increases. As soon as SNR_1 drops below the so-called Cell Search Threshold SNR_{CST} (point 1 in Figure 2), the MN enters the "cell-search" state and starts the scanning process to find better APs. In the scanning process, for every channel, the MN sends a Probe Request frame and waits for Probe Response frame from AP. Based on received Probe Response frames; the MN compares SNR values from scanned APs with the current one. The scanning process is repeated every Scanning Interval (T_{SI}) until one of scanned APs provide a better SNR than the current SNR by the amount of positive hysteresis Δ (point 4 in Figure 2). Now, the MN can switch to the channel used by this best AP, and start the re association process. In summary, the condition for the inter-AP handover is as follows

$$\begin{cases} SNR_1 < SNR_{CST} \\ SNR_2 > SNR_1 + \Delta \end{cases} (1)$$

The above handover algorithm reveals the main difference in handover procedures between WLAN and 3G cellular networks: in a 3G network, the MN can communicate simultaneously with two Base Stations (or Node Bs), and therefore a soft handover is possible, while in WLAN, MN has to perform hard handover which can only happen after a scanning cycle takes place. Our approach to handover, as described later in section 3.3, will be to identify the scanning cycle closest to the actual handover, and to transfer context information immediately after this scanning cycle is finished.

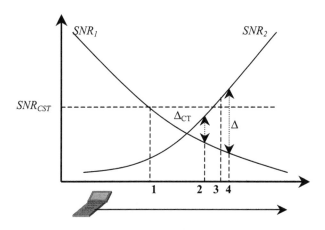

Figure 2 SNR Change between AP1 and AP2.

Overview of Context Transfer Protocol and Candidate Access Router Discovery Protocol

The Context Transfer Protocol (CTP) [7] is a fairly simple protocol that includes actions such as trigger - request – response. The protocol operation is illustrated in Figure 3. The protocol is initialized by either MN or AR depending on the CT trigger. The CT trigger is still an open issue as it depends on specific link layer technology. As shown later in section 3.1, our proactive scheme will use the condition from equation (3) as the CT trigger. In network-initialized scenarios, if the CT trigger is detected at the old AR, this AR will send the CT Data (CTD) to the new AR; otherwise the new AR will request the old AR to transfer context (CT Request). Upon receiving CTD, the new AR optionally may reply back to the old AR (CTDR – CT Data Reply). In both cases, the MN will send the CT Activation Request (CTAR). In mobile-initiated scenarios, the MN will send the CTAR upon receiving a CT trigger, usually from the link layer. Then, the new AR can request context transfer from the old AR.

There are several issues that arise when applying the CTP to specific services. For example, the CTP does not specify how dynamic context data such as Header Compression context can be transferred, as pointed out in [4]. More seriously, the CTP is insufficient in case of services involving network entities other than ARs. Intuitively, reestablishment of these services will require more time; hence reactive reestablishment may not be well suited to real-time applications. We will consider a proactive approach to context transfer in the next section

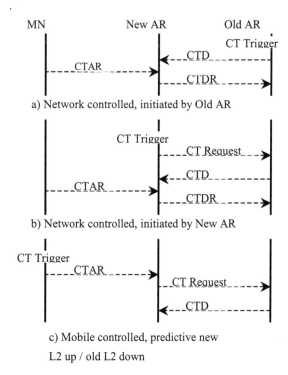

a) Network controlled, initiated by Old AR

b) Network controlled, initiated by New AR

c) Mobile controlled, predictive new

L2 up / old L2 down

Figure 3 The IETF Context Transfer Protocol Operation.

The Candidate Access Router Discovery (CARD) [9] is another draft resulting from the work of the IETF Seamoby WG. The objective of CARD is to identify (discover) the IP addresses of candidate ARs (CARs) for handover, and to discover their capabilities. Our proactive scheme will make use of the first CARD function mentioned above which, by CARD recommendations, can be implemented in centralized or decentralized schemes. The result of address mapping is included in the CARD Reply message that is sent back to the current AR. Reader should refer to [9] for more details of each option of mapping scheme.

As mentioned earlier, three protocols, Mobile IP, CTP and CARD are expected to work to together to facilitate seamless handover. Our contribution offers a way to combine these three protocols into a proactive handover and context transfer scheme. To ensure smooth operation of the proactive scheme, we also suggest the concept of forced handover. We will describe these proposals in detail in the next section.

3. PROPOSING SCHEME OF PROACTIVE CONTEXT TRANSFER IN FORCED HANDOVERS

In this section, we provide a method to estimate the time when the proactive context transfer should be triggered, and introduce the concept of forced handovers. Then, we give a detailed description of our proactive scheme for context transfer, and discuss aspects of its performance.

3.1 Estimation of Proactive Context Transfer Time

Recall the discussion in 2.1, our approach is to identify the best moment for proactive context transfer which is the time immediately following the scanning cycle closest to the actual handover. The procedure to identify this scanning cycle is described as follows. When in the cell-search state, after every scanning cycle, the MN estimates the time until handover as follows

$$T_{until\_handover} = \frac{\Delta - (SNR_2 - SNR_1)}{R_{SNR2} - R_{SNR1}} \quad (2)$$

where R_{SNR1} and R_{SNR2} are rates of SNR change for signals from the current AP and the scanned AP respectively. These rate values are obtained and updated on the basis of SNR measurements performed as part of the current and previous scanning cycles.

If the $T_{until\_handover}$ is less or equal to the T_{SI} (point 3 in Figure 2), the current scanning cycle is likely to be the second last (now called scanning-to-CT), and in the next scanning cycle (now called scanning-to-handover), the handover condition is likely to be satisfied. In short, the MN identifies the scanning-to-CT by

$$T_{until\_handover} \leq T_{SI} \quad (3)$$

To reduce computation, the MN may start to estimate the $T_{until\_handover}$ when the following condition is satisfied

$$\begin{cases} SNR_1 < SNR_{CRT} \\ SNR_2 > SNR_1 + \Delta_{CT} \end{cases} \quad (4)$$

where Δ_{CT} is less than Δ.

Δ_{CT} (point 2 in Figure 2) should be selected such that there is at least one scanning cycle before scanning-to-handover; therefore it can be defined from the following formula

$$\frac{\Delta - \Delta_{CT}}{R_{SNR2\,max} - R_{SNR1\,max}} = T_{SI} \quad (5)$$

where $R_{SNR1max}$ and $R_{SNR2max}$ are maximum rates of SNR change from the current AP and the scanned AP. The rate values of interest can be learnt (estimated) from previous measurements, or pre-set.

3.2 Forced or Active Handover

The above technique can produce a good estimate of the time for proactive CT (scanning-to-CT) and the time for handover (scanning-to-handover). This will be later confirmed by simulation. However, there is not a 100% guarantee that the handover condition (1) will be satisfied at the time of scanning-to-handover. One can argue that if the handover condition (1) is not satisfied at the time of scanning-to-handover, the MN may wait until the next scanning. However, in this case we need to set up a longer waiting time for the transferred context at the new access router, and consequently there may be more resources wasted. We suggest a forced handover, i.e. **the MN will make the handover after the scanning-to-handover time is reached, regardless of whether the handover condition (1) is satisfied or not.** The main advantage of such forced handover is that the MN knows exactly when the handover will happen, and therefore can set up an appropriate waiting time for the transferred context at the new access network. The forced handover at the link layer level also allows the MN sufficient time to prepare for the IP level handover. For example, the MN may use the forced handover as a trigger to send an Agent Solicitation message [2] to acquire the Router Advertisement message [2] from a new Foreign Agent (FA); therefore it can reduce the Agent Discovery time (T_{AD}) to as low as one Round Trip Time (RTT) between the MN and the new FA. The T_{AD} can be further reduced if the MN is able to receive the information needed for registration with the new FA through the CARD Reply message (more details will be given in section 3.3). This information enables the MN to create a Registration Request message [2] in advance and send it immediately after the MN re-establishes the link level connection to the new access network. The shortcoming of forced handover is that in some cases, handover is forced to happen when the handover condition (1) is not yet satisfied; therefore, the number of unnecessary handovers may increase. However, this number can be kept at a reasonably low level as shown in section 4.2.

3.3 Description of the Proactive Process

Now we will describe the proactive scheme for context transfer. Assume that MN is moving into an area where the SNR from the current AP drops below the SNR_{CST}, as illustrated in Figure 4.

(i) The MN starts a scanning cycle every T_{SI} seconds until the condition (4) is satisfied.

(ii) The MN starts estimation of the $T_{until\_handover}$ and continues scanning cycles until at least one of the scanned APs satisfies $T_{until\_handover} \leq T_{SI}$.

(iii) The MN collects L2 addresses of scanned APs satisfying the condition (3) (now we call them target APs), and sends them to the current AR via a CARD Request message.

(iv) Upon reception of the CARD Request message, the current AR resolves address mapping as described in the previous subsection and learns whether the MN is expected to perform an inter-AP or inter-AR handover. If the expected handover is inter-AP, the AR instructs the current AP to send the L2 context information to the new AP as specified in the IEEE 802.11 Inter-Access Point Protocol [5]; otherwise, the AR sends the CT Data message to the neighbor ARs. In case of the inter-AR handover, the current AR may ask the new AR to provide information necessary for the MN's registration with the new FA. Normally, this information is available via the Agent Advertisement message [2] broadcast periodically by Foreign Agents. Now, the current AR informs the MN about the result, including the handover type, the selected AP, the selected AR, and the registration information (if inter-AR handover) via the CARD Reply message.

(v) In the next scanning cycle, the MN performs handover at the link layer level to the AP specified in the CARD Reply message. If the inter-AR handover and the registration information are specified in the CARD Reply message, the MN creates a Registration Request message [2] and sends it to the FA immediately upon the reestablishment of link layer connectivity.

Now, assume that the handover is inter-AR

(vi) When the MN gets connected to the new AR, it sends the CTAR to activate the transferred context at the new AR.

(vii) Upon receiving the CTAR, the new AR starts context reestablishment at the new access network and sends the CTDR to the current AR (now the old AR).

(viii) Upon receiving the CTDR, the old AR takes appropriate action to delete old context.

In the step (iv), there may be a number of different scenarios possible; depending on how many APs satisfy the condition (3). We summarize these scenarios in the Table 1.

Table 1 Scenarios of inter-AP handovers and inter-AR handovers

Number of target APs	Scenarios
One	Both target AP and the current AP belong to the current AR → Inter-AP handover. Otherwise, inter-AR handover.
More than two	All target APs and the current AP belong to the current AR → inter-AP handover. Here, the current AR needs to make a decision which AP the MN should re-associate with. The decision may be based on communication link quality and /or current load at APs if this information is available.
	At lest one target AP connects to neighbor AR, while other target APs and the current AP belong to the current AR. As an inter-AP handover is normally preferred over an inter-AR one, the AR should select the AP such that only inter-AP handover is required.
	All target APs connect to an AR different from the current AR → Inter-AR handover.

3.4 Performance Metrics

While the advantages of forced handovers are very clear, in this paper we also investigate the disadvantages of the proposed scheme, and attempt to assess their significance. To evaluate the proposed proactive scheme, we define a **perfect handover** as follows:

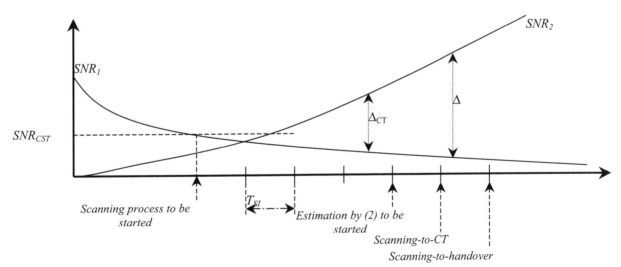

Figure 4 Time diagram of the proactive CT scheme.

65

(a) At the time of the scanning-to-CT, i.e. $T_{until\_handover}$ is less than or equal to T_{SI}, the condition (1) is not yet satisfied.

(b) The handover condition (1) is satisfied at the time of the next scanning, the scanning-to-handover.

(c) There is sufficient time for the proactive scheme to be completed.

Figure 5 shows an example of perfect handover. We define the proactive scheme completion as a state in which the MN is able to send the CARD Request message and receive the CARD Reply message. We denote the completion time as $T_{completion}$. The violation of any of the three conditions used in the above definition of perfect handover means that the handover is imperfect.

The imperfect handover of type I (Figure 6) can be considered a "premature" handover that could become an unnecessary handover. Therefore, the probability of this type of imperfect handover ($P_{im\_HO\_typeI}$) determines the upper bound on the probability of unnecessary handover (P_{un_HO}). The unnecessary handover happens when the MN moves in such a way that the handover condition (1) is never satisfied. For example, the MN changes the direction of movement or stops after the scanning-to-CT time. The unnecessary handover wastes network resources; therefore its probability should be as small as possible.

In contrast to the type I imperfect handover, the type II imperfect handover (Figure 7) can be seen as a "late" handover. The main concern with the late handover is that there is no opportunity for the MN to complete the proactive scheme as explained below. Firstly, the handover condition (1) satisfied at the time of the scanning-to-CT means that the MN might switch immediately to another channel i.e. another AP; consequently, the MN cannot send the CARD Request message to the current AR to start the proactive scheme. Secondly, assuming that the MN is able to postpone the handover until the next scanning cycle, there is still a possibility that the proactive scheme cannot be completed before the MN loses the current connectivity or re-associates with a new AP. Therefore, it is interesting to investigate the probability of the proactive scheme completion ($P_{II\_PS\_completion}$). It is also noted that the proactive scheme may not be completed even when a handover is not the type II imperfect handover i.e. the handover condition (1) is not satisfied yet at the time of the scanning-to-CT.

The above problem with incomplete proactive scheme (we call type III imperfect handover as illustrated in Figure 8) has consequences as illustrated in the following scenario. Let us assume that the MN is able to send the CARD Request message, but unable to receive the CARD Reply message. In such scenario,

(i) If there is more than one target AP, the MN does not know which one to switch to. To deal with this problem, the context data has to be transferred to all target APs; therefore there will be resource wasted at those target APs to which the MN will not connect. To reduce the waste, the new AR or AP, upon reestablishment of connectivity with the MN, should notify the target APs to release "unused" context. If there is only one target AP, there is no problem of wasting resources. Therefore, we may be interested in finding the probability of having only one target AP (P_{one_AP}), and of having more than one target AP, for example two target APs (P_{two_APs}).

(ii) Upon link reestablishment, the MN has to send the Router Solicitation message in order to reduce the Agent Discovery time (T_{AD}), as it does not know the expected handover type (i.e. inter-AP or inter-AR). If the expected handover is inter-AP, sending of Router Solicitation message will waste bandwidth.

(iii) In the case of inter-AR handover, T_{AD} can be reduced only to the RTT between the MN and the new AR, not to zero, as explained earlier in subsection 3.2.

Table 2 summarizes the three types of imperfect handovers and the parameters of interest to performance evaluation.

Of the three types of imperfect handovers, the type I and type III imperfect handovers are more important than type II. The type I imperfect handover is directly related to unnecessary handover, and the type III reduces benefits from the proposed proactive scheme. The type II imperfect is less significant, because it only affects the probability of incomplete proactive scheme as explained in the next section.

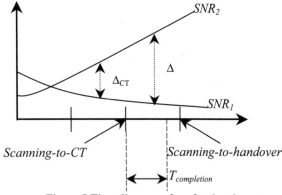

Figure 5 Time diagram of perfect handover.

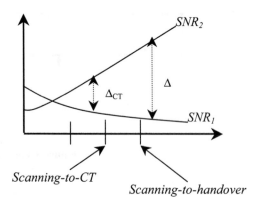

Figure 6 Time diagram of type I imperfect handover.

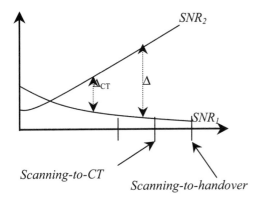
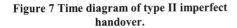

Figure 7 Time diagram of type II imperfect handover.

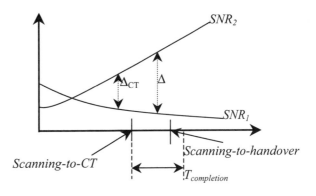

Figure 8 Time diagram of type III imperfect handover when the completion time is greater than the scanning time T_{SI}. (The type III imperfect handover also occurs if the MN loses the current connectivity before receiving the CARD Reply message.)

Table 2 Description of Imperfect Handover Types

Imperfect Handover Type	Description	Measured Parameters (Probabilities)
Type I ((a) is violated)	"Premature" handover as illustrated in Figure 6.	Probability of type I imperfect handover ($P_{im\_HO\_typeI}$). Probability of unnecessary handover (P_{un_HO}).
Type II ((b) is violated)	"Late" handover as illustrated in Figure 7.	Probability of type II imperfect handover ($P_{im\_HO\_typeII}$).
Type III ((c) is violated)	Incomplete proactive scheme as illustrated in Figure 8.	Probabilities of proactive scheme completion in two cases: (i) The handover is type II imperfect ($P_{II\_PS\_completion}$). (ii) The handover is NOT type II imperfect ($P_{notII\_PS\_completion}$). Probabilities of having one target AP (P_{one_AP}) and more than one target AP (for example, probability of having two target APs P_{two_APs}).

4. SIMULATION RESULTS

In this section, we describe the simulation scenario, present simulation results and follow up with a discussion.

4.1 Simulation Scenario and Objectives

The simulation area is covered by 61 APs distributed uniformly at a distance of 200m from each other. Transmission power of all APs is the same and in such level that there is no gap between coverage areas. The simulation area is assumed an open outdoor environment, therefore we can limit radio propagation model to path loss modeling. As future work we intend to investigate more complicated scenarios in semi-open or indoor environment i.e. we have to take into account fading channels or obstructions. Every AP, excluding the APs residing close to the edges, has 6 neighbor APs. In the simulation area, the MN is moving according to the random waypoint model as follows. After randomly selecting a destination, the MN moves towards the selected destination with a constant velocity v (the velocity v is randomly selected from a range of (0.5 m/s – 5 m/s)). After reaching the destination, the MN stops for the duration of pause time and then selects another destination and speed and moves again. The MN is always associated with an AP, and keeps monitoring SNR with this associated AP. As soon as this SNR drops bellows the threshold SNR_{CST}, the MN starts to follow the procedure described in section 3.3. Such scenario of the MN was repeated in a very large number of times to ensure that collected data are consistent.

Based on the discussion of performance metrics in the previous section, the simulation objective is to investigate

(i) How significant is the probability of unnecessary handover P_{un_HO}?

(ii) If type II imperfect handover occurs, what is the probability of proactive scheme completion ($P_{II\_PS\_completion}$)? This probability is also investigated for cases when the type II imperfect handover DOES NOT occur ($P_{notII\_PS\_completion}$), to complete the picture of proactive scheme completion.

(iii) What are the probabilities of having one target APs (P_{one_AP}) and two target APs (P_{two_APs})?

The above parameters are investigated in the context of different scanning intervals T_{SI} and hysteresis Δ. As we will see in the next subsection, smaller T_{SI} usually give a better performance. It is expected, since with smaller T_{SI}, the estimation of (2) is performed more frequently, hence produces more accurate predictions. On the other hand, smaller T_{SI} means that the MN has to interrupt the current communications more frequently in order to perform scanning.

4.2 Numerical Results and Discussions

The graphs in Figure 9 present probability of type I imperfect handovers ($P_{im\_HO\_typeI}$) and probability of unnecessary handovers (P_{un_HO}). As can be seen from the graphs, P_{un_HO} is always below the upper bound $P_{im\_HO\_typeI}$, and remains under 1% when the scanning interval T_{SI} = 2 sec, and under 1.5% when T_{SI} = 3 sec. As P_{un_HO} dramatically increases with the T_{SI} = 4 sec, the scanning interval should be selected to be no more than 3 sec. It is also observed that the P_{un_HO} is lower with smaller T_{SI}. The reason, as discussed earlier, is that more accurate estimation results from smaller T_{SI}. However, it is undesirable to select too small a T_{SI} that will significantly affect transmissions carried out by the MN.

The probabilities of type II imperfect handovers ($P_{im\_HO\_typeII}$) are shown in Figure 10. In the area of small Δ values, the $P_{im\_HO\_typeII}$ is high, but sharply decreases with the hysteresis Δ becoming larger. This is explained by the fact that it is to achieve the handover condition (1) with smaller Δ.

However, in the case of type II imperfect handovers, we are more interested in the probability of proactive scheme completion ($P_{II\_PS\_completion}$). Figure 11 and Figure 12 depict the $P_{II\_PS\_completion}$ when Δ = 1 dB and 7 dB, for two cases of scanning interval T_{SI} = 2 sec and 3 sec. We also obtained values of $P_{II\_PS\_completion}$ with other values of Δ between 1 and 7 dB. Those values fall within the two graphs included in the figures and were not included to improve the readability of the figures. For small Δ such as 1 dB, the $P_{II\_PS\_completion}$ is 100% as long as the completion time of the proactive scheme $T_{completion}$ is less than T_{SI}. Small Δ implies that handovers occur in the area well covered by the current AP; therefore the completion of proactive scheme may only be prevented by the sequence of events whereby by the time the current AR sends the CARD Reply message, the MN has already switched to the new AP.

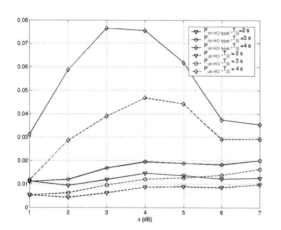

Figure 9 Probability of type I imperfect handovers ($P_{im\_HO\_typeI}$) & probability of unnecessary handovers (P_{un_HO}).

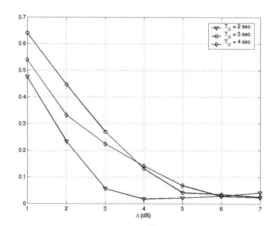

Figure 10 Probability of type II imperfect handovers ($P_{im\_HO\_typeII}$).

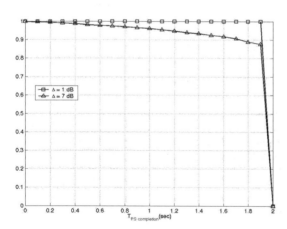

Figure 11 Probability of proactive scheme completion ($P_{II\_PS\_completion}$) in the case of the type II imperfect handover (scanning time T_{SI} = 2 sec).

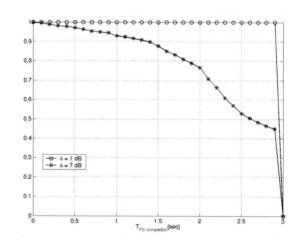

Figure 12 Probability of proactive scheme completion ($P_{II\_PS\_completion}$) in the case of type II imperfect handovers (scanning time T_{SI} = 3 sec).

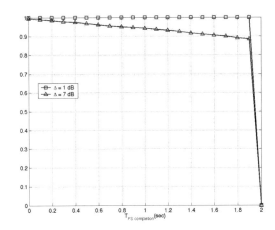

Figure 13 Probability of proactive scheme completion
$(P_{notII\_PS\_completion})$ **when the handover is NOT type II imperfect (scanning time T_{SI} = 2 sec).**

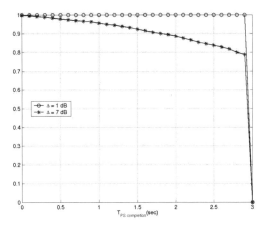

Figure 14 Probability of proactive scheme completion
$(P_{notII\_PS\_completion})$ **when the handover is NOT type II imperfect (scanning time T_{SI} = 3 sec).**

This observation leads to the following important remark: although small Δ implies higher $P_{im\_HO\_typeII}$, there is still a high probability that the proactive scheme will be completed. As Δ increases, handovers occur in the area closer to the boundary of AP's coverage; therefore the MN may miss the CARD Reply message not only because of switching to the new AP, but also because of losing current connectivity before switching to the new AR. As a result, the $P_{II\_PS\_completion}$ decreases with the increasing Δ, but can be still considered high. For example, if the proactive scheme can be completed within 1 sec, and Δ = 7 dB, the probability of proactive scheme completion is 95% and 92% with the T_{SI} of 2 s and 3 s respectively.

We also examined the probability of proactive scheme completion in cases where type II imperfect handover does not occur. From the results in Figure 13 and Figure 14, we observe very high $P_{notII\_PS\_completion}$ (100%) for small values of hysteresis Δ (1 dB) as long as the completion time of the proactive scheme $T_{PS\_completion}$ is kept less than the scanning interval T_{SI}. When the values of Δ become larger (e.g. 7 dB) i.e. the MN is closer to the boundary of AP's coverage, the $P_{notII\_PS\_completion}$ starts to decrease, but is still high. We also obtained the $P_{notII\_PS\_completion}$ with values of Δ between 1 and 7 dB. Those values fall within the two depicted graphs obtained for Δ = 1 dB and 7 dB, and are not shown in the figures for simplicity and clarity reasons.

By comparing $P_{notII\_PS\_completion}$ with $P_{II\_PS\_completion}$ for the same values of T_{SI} and Δ, we can see how the type II imperfect handover affects the probability of proactive scheme completion. For small Δ such as 1 dB, there is no difference between $P_{notII\_PS\_completion}$ and the $P_{II\_PS\_completion}$, and all probabilities are 100% if the proactive scheme is completed within the scanning interval T_{SI}. However, with larger Δ (7 dB), $P_{II\_PS\_completion}$ is slightly greater than $P_{notII\_PS\_completion}$ for short T_{SI} (2 sec), and the difference increases for longer T_{SI} (3 sec). For explanation, recall from the definition of type II imperfect handover that with the

bigger Δ the MN is closer to the coverage boundary in the case of "late" handover.

Now we turn our attention to the probability of a given number of target APs. Figure 15 and Figure 16 show probabilities of having one target AP (P_{one_AP}) and two target APs (P_{two_APs}), respectively. We note that these probabilities depend on positioning of APs in the simulation area. Firstly, we observe that higher P_{one_AP} (and consequently lower P_{two_APs}) results from larger value of hysteresis Δ. Larger Δ implies that the MN is likely to be close to only one target AP. Secondly, we also observe that higher P_{one_AP} result from smaller T_{SI}. However, the differences between the values of P_{one_AP} obtained with different T_{SI} are not great, and can be explained by the fact that smaller T_{SI} give more accurate estimation.

The simulation results and the discussion lead to the following preliminary conclusion. With appropriate scanning interval T_{SI}, the probability of unnecessary handovers can be kept as low as 1% or 1.5%. At the expense of some waste of resources resulting from the unnecessary handovers, the remaining majority of handovers can be very accurately predicted; therefore, the MN has sufficient time to prepare for the handover. The results also reveal that the main factor preventing the proactive scheme completion is switching to another AP too early, i.e. situation arising when the scanning interval T_{SI} is too short. In the simulations, the T_{SI} was selected from the range 2 - 3 s, and we believe that the proactive scheme can be completed within such scanning interval. Even when the proactive scheme fails to complete, i.e. when the MN is unable to receive the CARD Reply message, the proactive scheme still derives benefits, as context data is transferred to target APs and ARs. The disadvantage of not being able to complete the proactive scheme, as mentioned in the previous subsection, is that the MN does not know the expected handover type and, if there is more than one target AP, it also does not know which target AP to re-associate with.

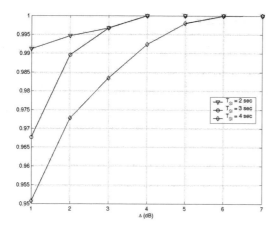

Figure 15 Probability of having one target AP (P_{one_AP}).

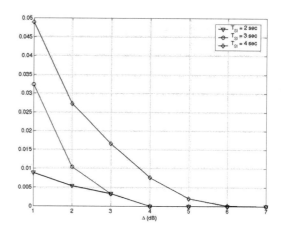

Figure 16 Probability of having two target APs (P_{two_APs}).

However, the second problem is insignificant because of the high probability of having one target AP (P_{one_AP}). In the case of having one target AP, the MN still knows the AP to switch to despite missing the CARD Request message. However, after re-association, the MN will still need to send a Router Solicitation message to discover whether it has moved to a new AR.

5. CONCLUSION & FUTURE WORK

In this paper, we presented a simple scheme for proactive Context Transfer in WLAN-based access networks. Based on observation of SNR changes, the proposed scheme predicts the best moment in time to perform the context transfer. In our scheme, the MN is forced to carry out a handover in the next scanning cycle. As the handover is forced to happen at a "planned" moment in time, the network can prepare for such event by selecting the best AP and AR, and transferring service context information between APs and ARs; therefore the scheme facilitates seamless mobility. Thanks to proactive handover scheme, the MN can significantly reduce the time of agent discovery; hence improve the handover latency. The improvement is achieved at the expense of small increase in the number of unnecessary handovers; this can be kept at a reasonably low level by appropriate selection of scanning interval.

We intend to carry the research described in the paper further. Firstly, we intend to verify the proactive scheme for other simulation scenarios, characterized by different radio propagation models. One of possible solutions is that the MN needs to keep track of target APs, so that it can eliminate impact of instant obstructions. Secondly, as pointed out in the discussion, the scanning interval is open to optimization: lower T_{SI} gives better performance of the proposed scheme, but affects (interrupts) the user communications more. Therefore, we need to investigate optimization of T_{SI}. One approach may be to use adaptive scanning interval. For example, initially T_{SI} can be selected large, and then be reduced adaptively as the MN approaches handover. Finally, we will investigate whether the proposed scheme can work in other types of access network, e.g. cellular network.

6. REFERENCES

[1] A. Mishra, M. Shin, and W. Arbaugh, "Context Caching using Neighbor Graphs for Fast Handoffs in Wireless Networks," Tech. Rep. University of Maryland, Computer Science Technical Report CS-TR-4477, 2003.

[2] C. Perkins (editor), "IP Mobility Support for IPv4," RFC 2002, IETF, Jan 2002.

[3] Cisco Systems Inc., "Cisco AVVID Wireless LAN Design Solution Reference Network Design", 2003.

[4] H.H. Duong, A. Dadej and S. Gordon, "Transferring Header Compression Context in Mobile IP Networks," Proc. of Inter. Symp. On Digital Signal Processing and Communication Systems, Coolangatta, Gold Coast Australia, pp. 223-228, 8-11 Dec 2003.

[5] IEEE, "Recommended Practice for Multi-Vendor Access Point Interoperability via an Inter-Access Point Protocol Across Distribution Systems Supporting IEEE 802.11 Operation, " *IEEE Draft 802.11f/D5*, Jan 2003.

[6] J. Kempf (editor), " Problem Description: Reasons for performing context transfers between nodes in an IP access network," RFC 3374, Sept 2002.

[7] J. Loughney (editor) et al., "Context Transfer Protocol," Internet draft (draft-ietf-seamoby-ctp-08.txt, work in progress), IETF, Jan 2004.

[8] Lucent Technologies Inc., "Roaming with WaveLAN/IEEE 802.11," Tech. Rep. WaveLAN Technical Bulletin 021/A, Dec 1998.

[9] M. Liebsch, A. Singh, et al., "Candidate Access Router Discovery," Internet draft (draft-ietf-seamoby-card-protocol-05.txt, work in progress), IETF, Nov 2003.

[10] T. Pagtzis, P. Kristein, S. Hailes and H. Afifi, "Proactive seamless mobility management for future IP radio access networks," *Computer Communication* vol. 26, pp. 1975-1989, 2003.

Secure Universal Mobility for Wireless Internet

Ashutosh Dutta
Tao Zhang
Sunil Madhani
Telcordia Technologies,
Piscataway, New Jersey

Kenichi Taniuchi
Kensaku Fujimoto
Yasuhiro Katsube
Yoshihiro Ohba
Toshiba America Research Inc.,
Piscataway, New Jersey

Henning Schulzrinne
Computer Science Department,
Columbia University, New York

ABSTRACT

The advent of the mobile wireless Internet has created the need for seamless and secure communication over heterogeneous access networks such as IEEE 802.11, WCDMA, cdma2000, and GPRS. An enterprise user desires to be reachable while outside one's enterprise networks and requires minimum interruption while ensuring that the signaling and data traffic is not compromised during one's movement within the enterprise and between enterprise and external networks. We describe the design, implementation and performance of a Secure Universal Mobility (SUM) architecture. It uses standard protocols, such as SIP and Mobile IP, to support mobility and uses standard virtual private network (VPN) technologies (e.g., IPsec) to support security (authentication and encryption). It uses pre-processing and make-before-break handoff techniques to achieve seamless mobility (i.e., with little interruption to users and user applications) across heterogeneous radio systems. It separates the handlings of initial mobility management and user application signaling messages from user application traffic so that VPNs can be established only when needed, thus reducing the interruptions to users.

Categories and Subject Descriptors

C.2.1 [**Computer-Communication Networks**]: Network Architecture and Design

General Terms

Management, Measurement, Performance, Experimentation, Security

Keywords: Mobility, Mobile IP, Hot Spot, 802.11, Handoff, Security

1. INTRODUCTION

A user should be able to roam seamlessly between heterogeneous radio systems, such as public cellular networks (e.g., GPRS, cdma2000, WCDMA networks), public wire-line networks, enterprise (private) wireless local area networks (LANs) and public wireless LAN hotspot networks that may use IEEE 802.11, Bluetooth, or other short-range radio technologies such as DSRC (Dedicated Short Range Communications). A user should be able to communicate and to access network services in a seamless and secure manner regardless of which type of access network one is using. A user should be able to maintain one's on-going secure application sessions when one moves across different access networks.

An enterprise user on an external network is typically required by one's enterprise to use a virtual private network (VPN) to connect into the enterprise network so that the enterprise network can authenticate the user and determine whether traffic from the user should be allowed to enter the enterprise network. The VPN also provides security protections to the user's traffic over an external network, which is usually an un-trusted network (e.g., any public network, the Internet, an Internet Service Provider's network, a corporate network other than the user's own corporate network). Such security protection may include integrity protection (i.e., protecting the information from unauthorized change) and confidentiality protection (i.e., preventing attackers on an external network from interpreting the contents in the packets).

Therefore, a key issue in supporting seamless and secure roaming across heterogeneous radio systems is how to meet the security requirements while a mobile is moving across enterprise networks and external networks and between different types of external networks.

Today's leading VPN technologies, such as IPsec [6], is a set of protocols defined by the Internet Engineering Task Force (IETF) to provide security protections such as authentication, privacy protection, and data integrity protection, do not have sufficient capabilities to support seamless mobility. For example,

- An IPsec tunnel will break when the mobile terminal changes its IP address as a result of moving from one network to another, unless proper measures in addition to the current standard IPsec is implemented. A new version

71

of IKE, IKEv2 [7], is supplemented with the mobility extension that may solve this mobility problem, but it is still in the early stages of development.

- VPN establishment may require user manual intervention when the user has to use a time-variant password to establish the VPN. This suggests that careful consideration needs to be given to when a VPN should be set up.

  o A VPN should be set up only when the user or a user application has a need for it. This suggests that user manual intervention will be incurred only when a user or a user application has a need for the VPN.

  o VPN setup should occur as infrequently as possible to reduce the frequency of user manual intervention and level of interruptions to user applications. This suggests that once a VPN is set up, it should be kept alive as long as allowed by the enterprise security policy. Since different enterprises will likely have different security policies, the VPN lifetime should be a flexible parameter.

- VPNs may introduce significant overhead. Many applications, e.g., non-confidential short messages from intranet to a mobile, may not need to be transported over a VPN. This suggests that on certain occasions the mobile should be allowed to receive non-confidential packets from the intranet without a VPN. Thus there is a need to have an architecture that can provide flexible VPN setup as on need basis.

In this paper, we propose a new architecture, named Secure Universal Mobility or SUM. It supports secure seamless mobility without requiring a mobile to maintain an always-on VPN. It does so by introducing another external home agent in the DMZ network of an enterprise in addition to the internal home agent. It can also incorporate more advanced mobility management techniques, such as MOBIKE, to reduce mobility management overhead. Alternatively it can allow application-layer protocols, such as SIP [12], to be used to support mobility. In addition it also involves heterogeneous access technologies such as 802.11 and CDMA1XRTT respectively.

The rest of the paper is organized as follows. Section 2 discusses some of the related work in this area. Section 3 describes the SUM architecture and its various associated functional components. Section 4 describes the test-bed set up and performance measurement citing the specific handoff scenario for both video and voice traffic that represent VBR (Variable Bit Rate) and CBR (Constant Bit Rate) traffic respectively. We conclude the paper in section 5.

2. Related Work

Over the past few years there have been several efforts to support seamless and secured mobility covering multiple administrative domains. Miu et al [13] describe architecture and systems that supports secured mobility between public and private networks. However it is limited to movement between similar kinds of networks e.g., (802.11b). Rodriguez et al [12] introduce the concept of mobile router where the end clients

with different access technologies connect to the mobile router's internal interface. In this case the end clients do not change their IP addresses rather the mobile router keeps on changing the external IP addresses as they move around and connect to different access networks such as GPRS, CDMA and 802.11b. Although it has taken care of technology diversity, network diversity and channel diversity to support variety of traffic, it has not discussed how to support security along with mobility. Snoeren et al [14] discuss fine-grained failover using connection migration mechanism. It achieves fine-gained, connection-level failover across multiple servers during an active session. However it does so by proposing changes in the TCP stack of the end clients without changing the application. References [15], [16], describe the integration of Mobile IP and IPSEC in an 802.11b environment but have not illustrated the use of heterogeneous access. Cheng et al [3] describe a novel approach that achieves smooth handoff during handoff, but it assumes foreign agents in the visited network and does not involve heterogeneous access technologies.

Recently, there has been much activity within the Internet Engineering Task Force (IETF) to develop solutions to maintain VPN connectivity while a mobile device changes its IP address. Adrangi et al [5] describe several scenarios of how a combination of Mobile IP (MIP) and VPN can support continuous security binding as a mobile device changes its IP address. However, it does not address how to support seamless handoff while preserving a VPN and also does not address heterogeneous access technologies. Luo et al [1] describe a secure mobility gateway that maintains mobility and security association between a mobile and a VPN gateway, but it does not offer flexible tunnel management techniques and has not explored mechanisms to provide smooth handoff. Birdstep (**www.birdstep.com**) proposes an approach that uses two instances of MIP to support seamless and secure mobility between an enterprise network and external networks. When a mobile moves to an external network, one instance of MIP is used to ensure that the VPN to the mobile does not break when the mobile changes its IP address. Another instance of MIP is used to ensure that packets sent to the mobile's enterprise network can be forwarded to the mobile through the VPN. A key advantage of the Birdstep approach is that it is based completely on existing IETF protocols. It, however, requires that a mobile keeps its VPN always on while the mobile is outside its enterprise network. Furthermore, it is limited to using MIP for mobility management. MOBIKE [2] provides an alternative approach to seamless mobility using the mobility extension of IKEv2 to support continuous VPN when a mobile changes its IP address. It thus avoids the need to use a separate mobility protocol, such as Mobile IP, to maintain continuous VPN and thus reduces the overheads needed to support secure mobility.

The proposed SUM architecture overcomes some of the limitations of the Birdstep approach when Mobile IP is used to support mobility. SUM uses the existing standard-based protocols such as IPSEC, Mobile IP, and SIP over transport layer mechanism (TCP, RTP/UDP) without any need to modify these. Make-before-break mechanism provides seamless mobility over heterogeneous access technologies.

3. The SUM Architecture

Figure 1 shows reference architecture without any related components that we will use to discuss SUM. An enterprise network is typically divided into intranets and de-militarized zones (DMZ). An intranet is the trusted portion of an enterprise network. A DMZ is a portion of an enterprise network that can be accessed from external networks under looser access control than the intranets.

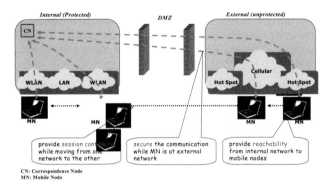

CN: Correspondence Node
MN: Mobile Node

Figure 1: Mobility scenario in DMZ-equipped enterprise network

Several modes of communication can take place with the mobile.

- Both the mobile and its correspondent hosts are inside the same enterprise network.

- The mobile host moves between its enterprise network and wide-area-based cellular networks while communicating with the correspondent host inside the enterprise network.

- The mobile in a cellular network communicates with correspondent nodes on external networks.

- The correspondent host within an enterprise can initiate communication with the mobile which is already in the cellular network.

- The mobile can initiate communication with a correspondent host which is within an enterprise.

Although a mobile can always maintain multiple simultaneous network connections, each over a different type of network, this paper focuses on the case where a mobile uses one interface (network) at any given time even if both the interfaces are up.

SUM seeks to achieve the following main capabilities:

- Maintain *reachability* from the intranet to a mobile user outside one's enterprise network in a secure manner with minimum interruption to the user and user applications.

   Reachability can be achieved using either network-layer or application-layer mobility management techniques. At the network layer, a mobile can have a permanent IP address for correspondent nodes to use to address their packets to the mobile regardless whether the mobile is inside or outside its enterprise network. This can be accomplished using, for example, MIP. In this case, a mobile relies on a MIP home agent (HA) in its mobility service provider's network (e.g., the enterprise network) to maintain the association between the mobile's home address and its current care-of address (a process commonly referred to as binding). When the mobile changes its IP address, it registers the new IP address with its home agent and the home agent will forward future packets to the mobile's new location. When a mobile moves onto an external network, a dual-HA approach can be used to maintain reachability from the intranet to the mobile. This will be discussed in more detail in Section 3.1.

   Application-layer protocols such as SIP may also be used to maintain reachability to mobiles on external networks. In this case, mobile's home SIP proxy inside the mobile's enterprise network keeps the up-to-date mapping between the mobile's application-layer address (e.g., SIP URI) and its current contact address. This will be discussed in more detail in Section 3.2.

- Provide a *security* environment to a mobile that is comparable to the security level the user gets inside his/her enterprise network, regardless of where the mobile is.

   Signaling messages and user application traffic can be protected using security mechanisms at different protocol layers. For example, IPsec [6] provides IP-layer encryption and authentication. Using IPsec-based VPNs to access an enterprise network, a user first establishes an IPsec tunnel to an IPsec gateway which typically resides in the enterprise's DMZ. The user-end of the IPsec tunnel is identified by the mobile's current care-of IP address that the user obtained from the visited network to send and receive IP packets over the visited network. This means that the IPsec tunnel will break when the mobile changes its care-of address as a result of moving from one network to another. Establishing a new IPSec tunnel requires several message (e.g., IKE messages) exchanges between the mobile and the IPsec gateway and can add excessive delay to the handoff. This can give rise to transient data loss when the mobile changes its IP address rapidly.

- Maintain *session continuity* as the mobile is on the move.

   A mobile can have various scopes of mobility such as micro mobility where only layer-2 network association may change, macro mobility where IP-layer network association changes, and domain mobility where a mobile moves from one network domain to another that may be operated by a different network provider. We have experimented with Mobile-IP and SIP-based approaches to support session continuity for the later two cases, as IP address does not change for micro-mobility case.

   To support seamless mobility, it is important to reduce handoff delay and transient data loss during handoff. Setting up VPN tunnels or establishing connectivity to a cellular data network (e.g., GPRS, WCDMA, or cdma2000) could introduce excessive delays that are intolerable to real-time applications. We will describe handoff processing and make-before-break handoff mechanisms that can significantly reduce handoff delay and data loss during handoff.

3.1 Mobile IP Based SUM Architecture

Figure 2 illustrates a MIP-based SUM architecture. It uses two MIP home agents. An internal home agent (denoted by i-HA) inside the intranet supports mobility inside the intranet. The external home agent (denoted by x-HA) in the DMZ handles a mobile's mobility outside the enterprise and ensures that a VPN to the mobile does not break when the mobile changes its IP address. The i-HA and x-HA collectively ensure that packets received by the i-HA can be forwarded to the mobile currently on an external network.

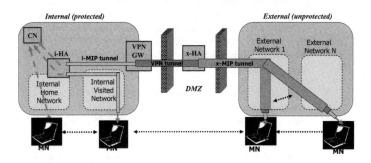

Figure 2: Secure mobility using MIP-based approach.

For clarity we define the following terms as these are used in the document.

- **i-HA-addr:** The IP address of the i-HA.

- **X**-HA-addr: The IP address of the x-HA.

- **i-MIP**: The instance of MIP used between a mobile and the mobile's i-HA.

- **x-MIP**: The instance of MIP used between a mobile and the x-HA.

- **i-CoA** (Internal Care-of Address): A mobile's care-of address registered with the mobile's i-HA.

- **x-CoA** (External Care-of Address): A mobile's care-of address registered with the mobile's x-HA.

- **VPN-GW**: VPN Gateway.

- **TIA**: Tunnel Inner address of the mobile

A mobile will have two MIP home addresses: an internal home address i-HoA in the mobile's internal home agent and an external home address x-HoA in the external home agent. The mobile's care-of address registered with its i-HA will be referred to as its internal care-of address and will be denoted by i-CoA. The mobile's care-of address registered with the x-HA will be referred to as its external care-of address and will be denoted by x-CoA. The instance of MIP running between the mobile and its i-HA is referred to as internal MIP or i-MIP. The instance of MIP running between a mobile and the x-HA will be referred to as external MIP or x-MIP.

When the mobile is within the intranet, standard MIP [8] or SIP mobility [9] can be used to support its mobility. In the rest of the paper, we focus on how to support mobility between enterprise network and external networks and mobility between external networks.

When a mobile moves into a cellular network, setting up the connection to a cellular network can take a long time. For example, we routinely experienced 10-15 second delays in setting up PPP connections to a commercial cdma2000 1xRTT network. In addition, establishing a VPN to the mobile's enterprise network could also lead to excessive delay. To enable seamless handoff, handoff delays need to be significantly reduced.

Therefore, we apply handoff pre-processing and make-before-break techniques to reduce handoff delay. In particular, a mobile anticipates the needs to move out of a currently used network, based on, for example, the signal qualities of the networks. When the mobile believes that it will soon need (or want) to switch onto a new network, it will start to prepare the connectivity to the target network while it still has good radio connectivity to the current network and the user traffic is still going over the current network. Such preparation may include, for example,

- Activating the target interface if the interface is not already on (e.g., a mobile may not keeps its cellular interface always on if it is charged by connection time),

- Obtaining IP address and other IP-layer configuration information (e.g., default router address) from the target network,

- Performing required authentication with the target network, and

- Establishing the network connections needed to communicate over the target network (e.g., PPP connection over a cdma2000 network).

Although both the interfaces are on at any specific point of time, the decision to switch over from one interface to another will depend upon a local policy that can be client-controlled or server-controlled. In this case the handover anticipation is purely based on signal-to-noise ratio (SNR) of the 802.11 interface. But this handoff decision could be based on any other specific cost factor.

When the mobile decides that it is time to switch its application traffic to the target interface, it takes the following main steps:

- Registers its new care-of address acquired from the target network with the x-HA.

- Establishes a VPN tunnel between its x-HoA and the VPN gateway inside the DMZ of its enterprise network.

- Registers the gateway end of the VPN tunnel address as its care-of address with the i-HA. This will cause the i-HA to tunnel packets sent to the mobile's home address to the VPN gateway, which will then tunnel the packets through VPN tunnel and the x-MIP tunnel to the mobile.

74

- When the mobile moves back to the enterprise network, the VPN and the MIP tunnels will be torn down. Dismantling the VPN tunnel may take up to few seconds, thus some of the packets which are already in the transit may get lost or may arrive at a later point. As a result, packets may arrive at the mobile out of order. Most of today's applications are capable of reordering of the out-of-sequence packets (e.g., out-of-sequence RTP packets).

When the mobile moves to another external network and acquires a new local care-of address (x-CoA), the mobile's x-HoA remains the same. Therefore, the mobile's VPN does not break. The mobile only needs to register its new local care-of address with the x-HA so that the x-HA will tunnel the VPN packets to the mobile's new location. Figure 3 illustrates how the MIP and VPN tunnels are set up during the mobile's movement from an enterprise network to an external network. If the mobile uses reverse tunneling, the data from the mobile will flow to the correspondent host in the reverse direction of the path shown in Figure 3.

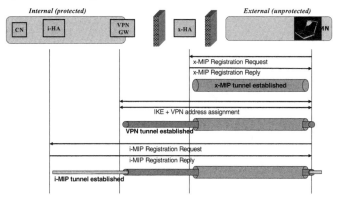

Figure 3: Interaction of Protocols for SUM using Mobile IP based architecture

Dynamic VPN establishment: Existing approach to secure mobility typically requires a mobile to maintain an always-on VPN when the mobile is outside its enterprise network regardless of whether the mobile has user traffic or not. This adds additional overhead on the mobile because of the tunnel keep-alive messages and may also introduce extra security risks (e.g., when the mobile device is lost). The proposed SUM solution employs a dynamic VPN establishment mechanism so that a mobile outside the enterprise network no longer needs to maintain a VPN to its enterprise network all the time. Instead, it establishes a VPN only when it needs to communicate with a correspondent host inside the enterprise network or to communicate through the enterprise network with a correspondent host on external networks.

Dynamic VPN establishment can be implemented using pre-condition features of SIP-based signaling when the correspondent host and the mobile are SIP-enabled. When the mobile is outside the enterprise network and has no user traffic to send into the enterprise network, it sets up only the two Mobile IP (MIP) tunnels: one from the i-HA to the x-HA and another from the x-HA to the mobile. This ensures that the SIP signaling from the correspondent host can reach the mobile and vice-versa. At this stage, no VPN is established, thus initial signaling to and from the mobile is not protected by a VPN

tunnel. Other security measures can be used to secure the initial signaling messages. For example, S/MIME [10] or TLS [11] (Transport Layer Security) could be used to secure the initial SIP signaling messages.

When a correspondent host (CH) wants to initiate communication with the mobile, it sends a SIP INVITE message to the mobile. This INVITE message can be sent in two ways. If the SIP proxy only has the home address of the mobile in the database, it will reply with a 302 redirect message in response to INVITE from the CH. CH will then send the INVITE to the internal home address of the mobile. The i-HA intercepts that packet, tunnels the packet to the x-HA which further tunnels the packet to the mobile. The SIP INVITE message notifies the mobile about an impending call. To answer this call, the mobile first checks to see if there is already an existing VPN. In the absence of a VPN, the mobile uses IKE to establish a new VPN. Then, the mobile uses the VPN to register the VPN gateway end of the VPN tunnel address with the i-HA so that the i-HA will forward future packets to the VPN gateway. At this point, the traffic between the mobile and the corresponding will travel through the VPN. Thus SIP OK from the mobile is carried within the VPN tunnel. This of course delays the SIP signaling little bit in the beginning but ensures that further communication is protected. There is no need to set up the initial double MIP tunnel if the SIP proxy at home has the prior knowledge of the mobile's COA. In that case initial INVITE from CN does not need to be carried over a double tunnel.

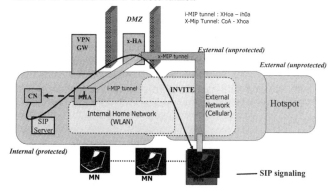

Figure 4: Dynamic tunnel management with SIP Signaling

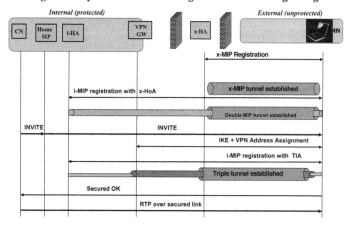

Figure 5: Protocol flow for dynamic tunnel management

Figure 5 shows the protocol flows associated with the dynamic tunneling mechanism using SIP's INVITE mechanism. Initially when the mobile is away it just sets up a double MIP tunnel, but the i-MIP tunnel is set up with a different care-of-address (x-HoA). After getting INVITE from the correspondent node original triple tunnels (x-MIP, VPN, i-MIP with TIA as the care of address) are set up so that the rest of the signaling and data traffic can flow over a secured link. In this specific scenario home SIP server keeps an association between URI and the home address of the mobile. Thus both the SIP server and home agents are used in this specific scenario compared to a complete SIP-based scenario described later on.

3.2 MOBIKE-based architecture

Using two instances of MIP and IPsec to support secure roaming introduces heavy tunneling overhead due to triple encapsulation (i-MIP, IPsec, and x-MIP) and may not be appropriate for bandwidth-constrained wireless networks. Techniques such as robust header compression (ROHC) [17] or IP-layer compression can reduce overhead. Most recently, mobility binding has been included as part of IKEv2 [7], where the mobile does not need to tear down its existing security association and re-establish it. Instead, it can modify the existing security association when it changes the IP address. This work is very recent within IETF and needs to be explored as part of future work.

As illustrated in Figure 6, using IKEv2, MOBIKE can modify the existing security association without re-establishing the IPsec tunnel when the mobile changes its IP address. This reduces header overhead by eliminating the need to use MIP (or another separate mobility management protocol) to maintain continuous VPN. MOBIKE-based architecture can work using both SIP and MIP-based mobility, where the mobile uses its VPN tunnel address (TIA) to register with the home proxy.

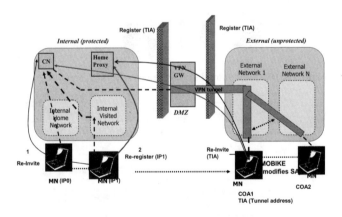

Figure 6: SUM architecture based on SIP and MOBIKE

4. Testbed Experimentation and Performance Evaluation

We have implemented the SUM framework in a multimedia test-bed as illustrated in Figure 7. It consists of an enterprise wireless LAN using IEEE 802.11b, CDMA1xRTT cellular network service from Verizon, and a public hotspot network

using IEEE 802.11 with EarthLink's DSL as backhaul into the Internet. But the measurements shown here reflect only moving back and forth between enterprise network and CDMA1XRTT access network and does not include the Earthlink hotspot.

We have realized both secure seamless mobility and dynamic VPN establishment techniques described earlier to reduce handoff delay and data loss during the handoff in both directions. Mobile-IP was used for mobility support and SIP was used to support dynamic VPN establishment. We have experimented with MIP HAs from SUN and Cisco. Nortel's VPN gateway is used to provide IPsec tunnels between a mobile on an external network and the enterprise network. A Microsoft Windows version of MIP client was used.

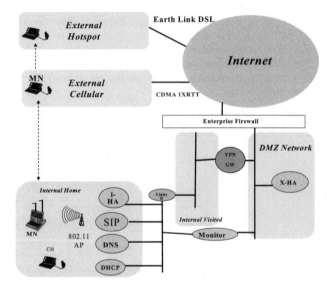

Figure 7: SUM Testbed

The mobile uses Sierra's PCMCIA card to access the cellular network. We have used both audio and video-based streaming applications using RAT and VIC tools respectively. The pre-processing and make-before-break handoff mechanisms establish the IPsec tunnel in the background while the communication is still going on through the 802.11 interface.

Table 1 shows the IP addresses associated with most of the functional components of the testbed. In this experiment the enterprise network is using private address space and thus we needed to install a Linux router in between that provides the NAT functionality. But in reality things would be simpler where the enterprise network also has globally routable IP address range. Both the internal home agent and external home agent are configured with a range of i-HoA and x-HoA addresses. These addresses get mapped to the corresponding i-CoA and x-CoA respectively. VPN gateway is configured with a set of TIA addresses. During the VPN set up with IKE, mobile node gets configured with a specific TIA address from this range. During the triple encapsulation process, TIA address of the mobile is sent as the care-of-address for setting up the i-MIP tunneling.

Table 1: IP address parameters

Entity	IP Address
CN	10.1.20.100
MN (i-CoA)	10.1.20.110 (DHCP)
MN (x-CoA)	166.157.173.122-(PPP assigned)
i-HoA	10.1.20.212
x-HoA	205.132.6.71
i-HA	10.1.20.2
x-HA	205.132.6.67
TIA (MN)	10.1.10.120
VPN-GW	205.132.6.66 10.1.10.100
SIP Server	10.1.20.3
DHCP Server	10.1.20.4

Figures 8, 9 and 10 describe the protocol flow sequences associated with the mobile's movement from the enterprise to the wide area network and then back.

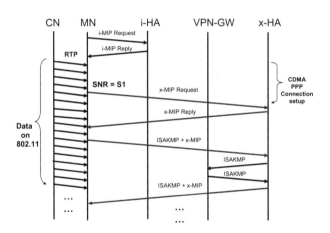

Figure 8: PPP setup over CDMA at SNR (S1)

Figure 8 shows the signal flows associated with the gradual movement of the mobile from the enterprise to the wide area network. Initially when the mobile is at the enterprise it receives the traffic sent by the CN (Correspondent Node) using standard ARP mechanism. Initially 802.11 is the only active interface and there is no PPP connection. As the mobile starts moving away, at a specific threshold value of Signal-to-Noise ratio (SNR = S1), PPP connection starts getting set up in the

background while mobile is still receiving traffic on its 802.11 interface.

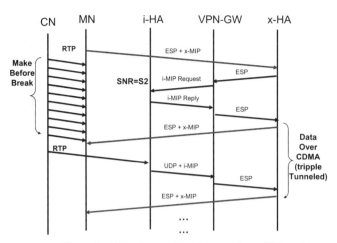

Figure 9: Make-before-break scenario at SNR = S2

Figure 9 shows the make-before-break situation as the mobile moves away further and loses connection with the 802.11 network. Make-before-break mechanism makes sure that the PPP connection and all the associated tunnels are set up before it loses the connection with the 802.11 network. Figure 10 shows the scenario as the mobile gets back home. VPN tunnel tear-down and CDMA disconnection take place in the background when the mobile still receives voice and video traffic in its 802.11 interface.

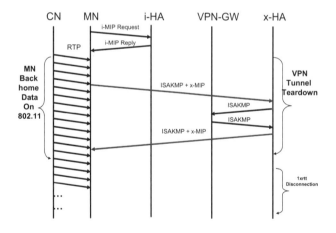

Figure 10: Mobile coming back home

We used ethereal and tcpdump measuring tool to capture the data on the mobile's multiple interfaces. These tools help capture the packets, the timing and their sequence numbers. Analysis from this log helps generate the performance parameters such as delay, packet loss, out of order packets etc. Figure 11 shows a set of measurements from the test-bed described in figure 7. It shows a sequence of the protocols executed on the mobile as it makes a transition from the enterprise network to the cellular network and then moves back. From several experiments, we observed that no packet was dropped during the mobile's movement from 802.11 networks to the cellular network, thanks to the handoff pre-processing and make-before-break handoff techniques. However there is at least one gap of about 500 ms between the last packet received on the 802.11 interface and first packet received on cellular interface. This delay is due to the i-MIP registration over the dual tunnel (x-MIP and VPN tunnel). The cellular network provided a lower data rate than the wireless LANs thus there is the low gradient.

As the mobile moves on to the next external network, it will simply update the x-HA with its new local CoA and does not need to re-establish the VPN. When the mobile moves back to its home network inside the enterprise network, some packets received by the mobile were out of sequence. This was due to the fact that the mobile already began to receive traffic from the enterprise network using its 802.11b interface while the VPN and MIP tunnels are being dismantled on the CDMA interface. According to the implementation it takes up to 5 seconds before the cellular interface is taken down after the mobile has registered its 802.11b interface with the internal home agent. During this time, the mobile continues to receive the transit traffic on its cellular interface, allowing the mobile to recover the transit packets which are already in the flight.

Figure 11 shows the relative sequence of protocol flow during a mobile's handoff from an 802.11b network to cellular and then back. We are showing only three types of protocols here in the diagram, protocol 1 denotes the RTP packets received on the mobile, protocol 2 shows the IPSEC instances, and protocol 3 denotes the Mobile IP signaling between the mobile, i-HA and x-HA. We set up an NTP server and synchronized the correspondent host and mobile node to measure the timing associated with each operation.

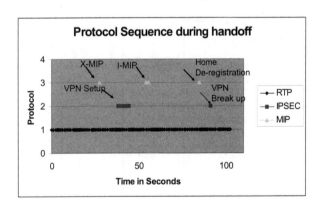

Figure 11: Protocol sequence during Handoff

From the measurements taken on the mobile, we observed that it takes about 10 seconds for the PPP negotiation to complete, about 300 ms for the x-MIP registration to complete, about 6 sec for VPN tunnel setup, 400 ms for the i-MIP registration, and 200 ms for mobile IP de-registration when the mobile is back.

Thus packet loss during all three types of movements (i.e., enterprise (802.11b) to the cellular network, cellular network to hotspot (802.11b), and hotspot (802.11b) to hotspot (802.11b)) are avoided using proactive movement detection handoff scheme. Because of the limited bandwidth on the cellular

network (60 kbps throughput) voice quality gets affected if proper codec was not chosen. We however plan to try this experiment on a high speed cellular network such as CDMA1XEVDO.

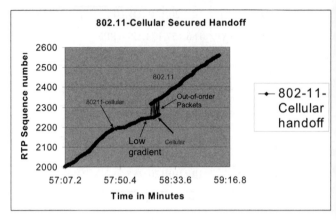

Figure 12: RTP packets at mobile during handoff

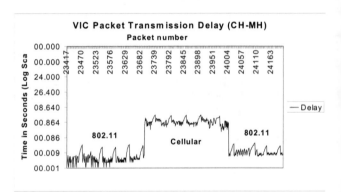

Figure 13 (a) Packet transmission delay

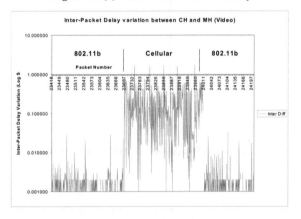

Figure 13 (b) Inter-packet departure and arrival variation delay for VBR (Video)

Figure 12 shows RTP sequence numbers received on the mobile as it performs handoff between 802.11b network and cellular network. As explained earlier, we observe that the packets are received out of sequence for about five seconds after the mobile has come back to the 802.11 network. While in cellular network "Low Gradient" of the curve in Figure 12 is due to the low bandwidth in the cellular network.

Figure 13 (a) and (b) show the transmission delay in logarithmic scale for the RTP packet from a streaming video being sent using VIC, and variation between inter-packet departure gap at CH and inter-packet arrival gap at MH. This variation delay seems to be more prominent in cellular network than 802.11b and thus will give rise to more jitter in the cellular network. It is interesting to note that video streaming is

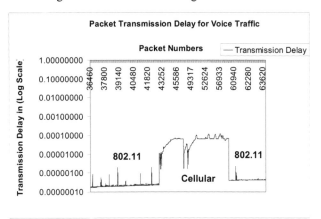

Figure 14 (a) Packet Transmission Delay

bit bursty in nature and thus has larger gap between bursts of packets (~1.5 s – 2.6 s) than the delay between consecutive packets within a burst (2 ms- 5 ms) both at the sending side and receiving side. Extent of burstiness of video traffic is affected by the frame rate of the video at the sending host. Figure 14 (a) and (b) show the results from the VoIP application using Robust Audio Tool (RAT). In the specific experiment we have used GSM encoding with a payload of 33 bytes. Compared to VIC-based video streaming, VoIP application is CBR (Constant Bit Rate) traffic and there is no burstiness. Transmission delays in Figure 10 (a) are in logarithmic scale.

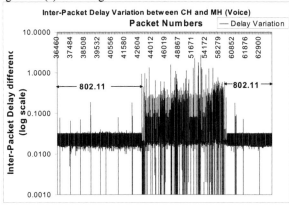

Figure 14 (b) Inter-packet departure and arrival delay variation for CBR (Voice)

As is observed in Figure 14, the transmission delay for RTP packet is almost 5 ms when the mobile is in 802.11b network but

transmission delay for the packet increases in a random fashion as it moves to the cellular network and then saturates. This could be attributed to the queuing delay in the network.

Figure 14 (b) shows variation delay between inter-packet departure gap at CH and inter-packet arrival gap at MH. This variation seems to be more prominent in cellular than in 802.11b network for voice traffic also and may affect the audio quality in the cellular network.

Inter packet delay at the sender will depend partly upon the codec type and unit of transmission packet. We also observed that home agent could not encapsulate many of VoIP packets during its movement to cellular network and while within cellular. However we did not lose any packet for VBR traffic such as video streaming traffic.

Table 2 shows the timings associated with PPP setup, packet transmission delay, inter-packet arrival delay both at sender and receiver, MIP registration, DHCP and VPN setup.

Table 2. Timing and CODEC details

Type of operation	Timing
PPP setup	10 sec
X-MIP	300 ms
VPN Tunnel setup	6 sec
I-MIP	400 ms
I-MIP at home	200 ms
IPSEC processing (end host)	60 ms
DHCP (address acquisition)	~ 3 sec in 802.11b
One way Transmission Delay	Video – (a) ~5 ms (802.11b), (b) 500 ms - 2.5 sec (CDMA) Audio - (a) ~ 4 ms (802.11b), (b) gradual increase and then saturates in (CDMA)
Inter-packet gap	VIC (VBR)– variable (intra-burst, Inter-burst) RAT (CBR)– 16 ms – 32 ms
CODEC	RAT–GSM, Silence suppression Off VIC - H.263, 50 kbps, 5 fps

An overall analysis of the results from the above prototype experiment shows that a make-before break technique adopted here help achieve smooth handoff while preserving the security of the data and signaling during the mobile's handoff. VBR-based video traffic is bursty in nature and thus gave a different set of values for inter-packet gap and transmission delay compared to a CBR based voice traffic.

Transmission delay in 802.11b network does not show that much variation as compared to the cellular network. This could be attributed to the fact that cellular medium is a shared one and is subject to bandwidth fluctuation and interference more than the 802.11 network which is under controlled environment. In a real life scenario enterprise network is subjected to other traffic and may attribute to the signaling delay and packet loss.

In this specific experiment, we have not included the interaction with AAA server during its movement from enterprise to cellular environment. In reality hotspot and cellular networks may belong to two different administrative domains and the user may have separate subscription profile. Thus the mobile will need to contact the AAA server to perform profile verification before being able to continue the communication with the correspondent host. As part of our future work we plan to build a Secure Mobility Gateway (SMG) that will have the prior arrangement with the AAA servers in each of the roaming domains and will work as a broker agent between the domains. By having a dual functionality (e.g., AAA broker and external home agent) the mobile does not need to communicate with two different AAA servers belonging to two different domains. Recently there has been proposal in the "Core Networks" working group of 3GPP2 [18] that suggest that AAA profile verification and Mobile IP authentication can also take place in parallel. This mechanism will make the handoff between administrative domains bit faster as the AAA profile verification can take place in parallel while Mobile IP authentication can take place using other access authentication protocols such as PANA [19] (Protocol for carrying Authentication to Network Access).

5. Conclusions

We have presented an architecture and test-bed realization of secured universal mobility across heterogeneous radio systems including 802.11b and CDMA-based networks. Both Mobile IP and SIP-based architecture were discussed. Security, mobility, reachability, and dynamic VPN tunnel management are some of the highlights of the architecture. Test-bed experiments show that smooth handoff is achieved during the movement between heterogeneous networks, but an additional delay was introduced during the transition from 802.11 network to another cellular network and while in cellular network. VoIP and video streaming traffic were used as CBR and VBR application respectively. Both of these applications showed different characteristics in terms of packet transmission delay, burstiness, jitter and packet loss for both the types of access networks during the handoff experiment. SIP and Mobike-based approaches seem to provide alternatives to MIP-based approaches and could reduce tunnel overheads.

6. References

[1] Hui Luo et al, "Integrating Wireless LAN and Cellular Data for the Enterprise", IEEE Internet Computing, April 2003

[2] T. Kivinen, "MOBIKE protocol", draft-kivinen-mobike-protocol-00.txt, Internet Engineering Task Force, Work in progress

[3] Ann-Tzung Cheng, et al , "Secure Transparent Mobile IP for Intelligent Transport System" ICNSC 2004, Taipei

[4] www.birdstep.com

[5] F. Adrangi, H. Levkowetz, Mobile IPv4 Traversal of VPN Gateways, <draft-ietf-mip4-vpn-problem-statement 02.txt>, Work in progress, IETF

[6] C. Kaufman et al, Internet Key Exchange (IKEv2) Protocol, draft-ietf-ipsec-ikev2-14.txt, IETF , work in progress

[7] C. Perkins et al, IP Mobility Support for IPv4, RFC 3344

[8] H. Schulzrinne, Elin Wedlund, "Application Layer Mobility using SIP" ACM Mobile Computing and communications Review, vol 4 no 3, p47-57, July 2000

[9] P. Hoffman et al, "S/MIME Version 3 Message Specification for S/MIME", RFC 2634, IETF

[10] T. Dierks et al, "Transport Layer Protocol Version 1.0", RFC 2246, IETF

[11] J. Rosenberg, H. Schulzrinne et al, "Session Initiation Protocol" RFC 3261, Internet Engineering Task Force

[12] P. Rodriguez, R.Chakravorty et al., "MAR: A commuter router Infrastructure for the Mobile Internet", Mobisys 2004

[13] A. Miu, P. Bahl, "Dynamic Host configuration for Managing between Public and Private Networks", USITS, 2001, San Francisco

[14] A. Snoeren, D. Andersen, H. Balakrishnan, " fine-grained Failover Using Connection Migration", USITS, 2001, San Francisco

[15] M. Barton, D. Atkins et al., "Integration of IP mobility and security for secure wireless communications", Proceeding of IEEE International Conference on Communications (ICC) 2002

[16] A.Dutta, S. Das et al, "Secured Mobile Multimedia Communication for Wireless Internet", ICNSC 2004, Taipei

[17] C. Bormann et al, "RObust Header Compression", RFC 3095, IETF

[18] 3rd Generation Partnership Project 2, www.3gpp2.org

[19] A. Yegin et al, "Protocol for Carrying Authentication for Network Access Requirements", work in progress, IETF PANA working group

Network Selection and Discovery of Service Information in Public WLAN Hotspots

Yui-Wah Lee *and Scott C. Miller
Bell Laboratories

ABSTRACT

In a public WLAN hotspot, a roaming mobile terminal (MT) may be within radio range of more than one access point (AP), each of which may or may not have roaming agreements with the service provider of the user of the MT. In this case, the MT may need to discover some *service information* before it can make an intelligent *network-selection* decision. The most critical is *roaming information*; while other information such as *security policies, price, AP workload* may also be useful. Currently, roaming information is typically provisioned on the MTs as static *roaming tables* or *roaming lists*. However, this approach may not scale well when there are millions of hotspots globally. Addressing this shortcoming, recently several solutions have been proposed by different groups. In this paper, we contrast these solutions, and propose our own solution called *Roaming Information Code (RIC)*, which can be transported as SSID or a new Information Element of the 802.11 standard. RIC is scalable and can be fully backward compatible with existing APs (if transported as SSID). Furthermore, it does not hinder fast handoffs. In the second half of the paper, we will also discuss two other schemes addressing the other service information: a scheme called *RIC-VAP* for provider-specific security information; and a scheme that allows an AP to announce price and workload information.

Categories and Subject Descriptors

[**C2.1 Network Architecture and Design**]: Wireless communication; [**C2.3 Network Operations**]: Public networks

General Terms

Standardization

---

*Corresponding Author. Address: Yui-Wah (Clement) Lee, Bell Laboratories, 101 Crawfords Corner Road, Holmdel NJ 07733 USA. Email: `leecy@bell-labs.com`. Tel:+1-732-949-4851. Fax:+1-732-949-7397

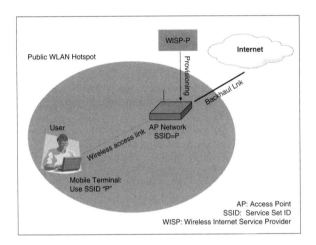

Figure 1: Public WLAN hotspot service

Keywords

Network Selection, Discovery of Service Information, Public WLAN Hotspots, Wi-Fi, Roaming

1. INTRODUCTION

This paper is about the emerging services for public wireless Internet access offered through technologies of wireless local-area network (WLAN), such as those compliant to the Wi-Fi or IEEE 802.11 standard [18] [15]. These services are also commonly known as "public WLAN hotspot services."

Figure 1 shows how such a service is typically rendered. A wireless Internet service provider (WISP) would set up a network of wireless access points (APs), with backhaul links into the Internet, in various public locations, such as coffee shops, airport lounges, hotel lobbies, etc. Its users will be carrying some WLAN-enabled mobile terminals (MTs), such as laptop computers, PDAs (personal digital assistants), or smartphones. Through these MTs, these users can get Internet access in these public WLAN hotspots.

Currently, the MT discovers an AP network based on its name. In the IEEE 802.11 standard[15], the name is a string and is also called *Service Set Identity*, or SSID. It is also commonly known as the "network name." It is not usually called "AP name" because multiple APs on the same AP network may use the same SSID. The procedures to use an AP network are as follows:

1. The APs are configured with the network name. They periodically broadcast a beacon, which contains various information including the network name.

2. The MT is configured with a list of network names. The list tells the MT which APs will offer Internet-access service to the user. (Users obtained these names from their WISPs.) The MT constantly listens to beacons broadcast by APs. It inspects and sees if the network name contained in these beacons matches one of the names in its list. If so, it will elect to use the AP network by performing a link-level "association" procedure with the AP. Note that user authentication typically will also happen immediately preceding or following the association procedure.

3. After a successful association of the MT and the AP (and also after a successful user authentication), the user can access the Internet through the AP network.

In a roaming situation, the MT has to choose an AP to associate with when it is within range of one or more APs. We call this problem *network selection*. Figure 2 depicts the problem. In this paper, we advocate that there has to be some support for the *discovery of service information* so that the MT can make an intelligent decision.

Various types of service information may be needed. At a minimum, the MT has to know whether the AP will offer it service. In this paper, we call this information *roaming information*. As explained in Section 3, a given AP may or may not have roaming agreements with the service provider of the user of the MT.

In addition to roaming information, other information may also be needed. For example, the MT may need to know the security policies, price information, workload information, etc. of the AP network.

Currently, roaming information is typically provisioned on the MTs as static *roaming tables* or *roaming lists*. However, this approach may not scale well when there are millions of hotspots globally. Addressing this shortcoming, recently several solutions have been proposed by different groups [8, 2]. In this paper, we contrast these solutions, and propose our own solution, *Roaming Information Code (RIC)*, which can be transported as SSID or a new Information Element of the 802.11 standard. RIC is scalable and can be fully backward compatible (if transported as SSID). Furthermore, it does not hinder fast handoffs. In the second half of the paper, we will also discuss two other schemes addressing the other service information: a scheme called *RIC-VAP* for provider-specific security information and a scheme that allows an AP to announce price and workload information.

This paper will be structured as follows. In the next section, we will begin with a more detailed description of the related portions of the IEEE 802.11 standard, since they affect the design of a backward compatible solution. Then the paper is divided into two halves. In Section 3 to 5, we will be focusing on roaming information. In Sections 6 and 7, we will be focusing on the other service information. Finally, we will briefly contrast our work with other related work in Section 8, and then conclude in Section 9.

2. BACKGROUND

The issues of network selection and discovery of service information are common across different types of WLAN

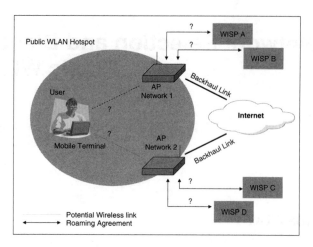

Figure 2: Selecting an AP network in a public wireless hotspot

technologies. However, for easier discussion, in this paper, we will be referring to one specific technology — the one that is defined in the IEEE 802.11 standard [15]. The ideas and principles presented in this paper may also be applicable to other WLAN technologies, such as HyperLAN.

In the 802.11 standard, there are three types of frames: Data, Control, and Management. Management frames are of particular relevance to this discussion, and some of the important management frames are *Beacons, Probe Request, Probe Response, Association Request, Association Response, Dis-association,* etc.

Information is carried in these frames as *headers, Fixed Fields (FFs)* (fixed-length mandatory frame-body components), and *Information Elements (IEs)* (variable length mandatory and optional frame-body components).

SSID, or *Service Set IDentity*, is an important IE carried in management frames. Its payload can have zero to 32 octets, and it is a configurable string identifying the AP and its network. (There are some examples in Figure 3). Normally an AP will broadcast the SSID in Beacons periodically (say, every 60 ms). However, an MT can also send a Probe Request with an embedded SSID IE to invite an response from an AP. In this case, an AP configured with the same SSID will reply with a Probe Response that contains not only the SSID but also other useful FFs or IEs, such as Beacon Interval, Capability Information, and Supported Rates. SSID is also sent in all Association Requests.

The standard does not specify how many SSIDs can be put into a beacon, nor does it specify whether a given AP can broadcast different SSIDs in different beacons. The common practice has been that each AP has just a single SSID. However, this has changed recently, and there are vendors making AP products that broadcast multiple SSIDs. Unfortunately, without a standard these products may not be inter-operable with the products of other vendors. We will discuss more details about these in Section 4.2.

SSID should not be confused with another important ID in the standard — the *BSSID*, or *Basic Service Set IDentity*. BSSID is a 48-bit (6-octet) field that use the same format as an IEEE 802 MAC address. In the infrastructure mode, the BSSID is the MAC address of the AP. BSSID is sent in all data, control, and management frames.

Now let us turn to another important aspect of WLAN: security. The original security scheme in 802.11 is called WEP, for Wired Equivalent Privacy [15]. However, this scheme has been proven to be cryptographically weak [7, 9], so there has been an effort to define a new and more secure scheme. This is done in the Task Group I of the IEEE 802.11 Working Group. The standardization process has been lengthy – not until Jun 2004 was the 802.11i standard officially approved by IEEE. As an interim measure, in October 2002 the industry consortium Wi-Fi Alliance decided to endorse a subset of the standard and to promote the use of this subset in the industry. The result was the security specification called WPA (Wi-Fi Protected Access)[5].

Besides all these link-layer security schemes, another security scheme is also commonly adopted in the public hotspot industry. This is called the Universal Access Method (UAM) [6]. In fact, this is the most commonly used security scheme in the industry. In this scheme, the AP will set to run as an open system (i.e. no link-layer authentication and encryption). An MT can freely associate with an AP, and the WISP asserts control on the IP level. Typically when the user first accesses a web page, her HTTP request will be redirected to a web server, through which the user can perform an web-based authentication procedure. Only after a successful authentication can the user access the Internet —- before that all the user's packets to and from the Internet will be dropped. Although by running as an open system, the AP does not offer link-layer data encryption, a user may still protect her privacy by running encryption on the IP level using technologies such as IPSec.

How does an AP announce the desired security scheme that it wants to use? There is an IE (information element) called Capability Information sent in Beacons and Probe Responses. In the IE there is a Privacy field (Bit 4). An AP sets the field to 1 if it requires the MT to use WEP; it sets the field to 0 if it does not. If an AP requires the MT to use WPA or 802.11i, it will use a new IE called RSN (Robust Security Network) to announce the desired authentication and cipher suites.

3. DISCOVERY OF ROAMING INFORMATION

In Figure 3, we tabulate some of the popular WISPs and the SSIDs / network names that they use for their AP networks. Typically a WISP will tell its users to configure only its own network name into their MTs. So these users will get service from only the hotspots of the WISP (i.e. no roaming). A notable exception is Boingo, which actually gives a table of network names to its users (more on this in Section 4.1).

WLAN technologies use short-range radio, so by nature a WLAN hotspot can cover only a small local area, with a radius of about 100 meters. Therefore, to offer a large coverage, a WISP needs to provision many hotspots. It can achieve this by either building as many as possible of its own WLAN hotspots, or it can achieve this by allying with other WISPs. By signing some *roaming agreements* with these other WISPs, it can let its users get services when they roam into the WLAN hotspots of these WISPs. The general feeling of the industry is that roaming support is very critical

Wireless ISP	Network name
T-Mobile	tmobile
Verizon Wireless	Verizon Wi-Fi
Starwood hotels	Turbonet
Cometa	Cometa-Hotspot
Wayport	Wayport_Access
Surf and Sip	SurfandSip
Fatport	fatport
Boingo	*various ...*

Figure 3: Examples of WISPs and their chosen network names / SSIDs. Boingo does not use a single SSID but use a roaming table (see Section 4.1)

to the widespread acceptance of the public WLAN hotspot services.

The roaming agreements can be *bilateral* or *multilateral*. In the former case, a WISP signs up pairwise network-sharing roaming agreements with other WISPs individually. In the latter case, a WISP joins a *roaming organization*, through which it has a one-to-many roaming agreements with all the other members of the organization.

The multilateral approach will scale better with respect to the number of WISPs. For N WISPs to have full roaming relationship with each other, in the bilateral model, $N(N - 1)/2$ bilateral agreements will be needed; whereas in the multilateral model, only N multilateral roaming agreements will be needed (Figure 4). The multilateral model facilitates not only the forming of roaming relationship but also the accounting, billing, and marketing of the services.

In either model, when a user roams to a hotspot, her MT has to have some information about the roaming agreements to make an intelligent decision on network selection. The information can either be *statically provisioned* or *dynamically discovered*. We will discuss the various proposals in the next two sections. Here we outline some of the key requirements for a good solution. These will form the basis upon which we compare the various proposals.

Scalability. Potentially there can be millions of hotspots globally. The solution should be able to perform well at this scale.

Backward Compatibility. Ideally the solution can be used without requiring the WISPs or AP operators to upgrade the hardware or firmware of their existing equipment. Note, however, that there are already two classes of equipment in the market because of the recent introduction of the WPA specification [5]: some support 802.1x but some do not. We will make a distinction to these two classes when we discuss backward compatibility.

Does not require technology changes. Some of the solution will require technology changes that typically will have to go through the lengthy process of technology standardization. Ideally the solution does not require technology changes.

Does not require association before discovery of roaming information. As we can see in Section 4, some of the solutions require the MT to first associate with an AP *before* the MT can acquire the in-

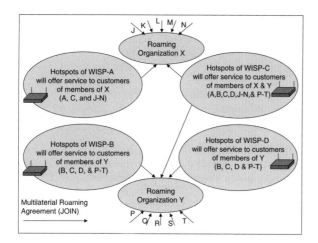

Figure 4: WISPs forming multilateral roaming agreements with others

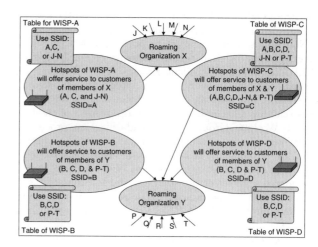

Figure 5: Using roaming tables to aid network selection

formation about the AP. We consider this requirement unfavorable. Because this requirement means that the MT cannot discover the information by simply listening to the broadcast beacons or by the probe-request/response mechanism. Furthermore, if the MT is currently associated with another AP, then the MT has to first *dis-associate* with the current AP before it can acquire information about a candidate AP. This will hinder the implementation of fast handoff.

Does not hinder fast handoff. When supporting multimedia applications, such as VoIP or streaming audio/video, an MT has to complete an *inter-AP handoff* with a very small latency (e.g. less than 50 ms). Otherwise, the users' delay-sensitive multimedia streams would be disrupted. Although in this paper we will not be discussing the exact mechanism of fast handoff, we will point out the problems if some solution will hinder the implementation of fast handoffs. Examples of these problems include the need of association before discovery (previous bullet point) and the need of manual intervention during handoffs.

4. EXISTING SOLUTIONS

4.1 Roaming Table

A straightforward solution for managing the roaming information is to statically provision the information in a *roaming table* or *roaming list*[1] on the MT. In this solution, each WISP will pick a network name (SSID) identifying itself, and it will configure all its APs to broadcast that network name. It will also configure a roaming table on each MT, listing all the network names of its roaming partners. Using the roaming situations illustrated in Figure 4, the users of these WISPs will be carrying roaming tables as illustrated in Figure 5.

---

[1]In implementation, the data structure is more likely a table than a list. This is because other information such as addresses of hotspot and support telephone numbers may also be contained in the data structure.

A WISP, Boingo, is using this approach. It provides its users a client software to be installed on their Windows-based laptops or their Pocket-PC-based PDAs. The client contains a table listing all the network names of Boingo's partners.

The advantage of this approach is that it is fully backward compatible with the existing APs (802.1x and the earlier non-802.1x ones). It does not require an MT to first associate with an AP before discovery of roaming information. There is also no obvious hindrance that prevents the implementation of fast handoffs.

However, there is a scalability issue with respect to the number of network names globally. Currently, the common roaming tables have less than a few thousands entries. However, as WLAN hotspot becomes more and more popular, the number of hotspots globally may grow into the range of millions or tens of millions.

To have a very rough estimate of the potential number of hotspots globally, perhaps we can look at some other infrastructures as references. In the banking industry, it was reported that there are more than 900,000 ATMs (Automatic Teller Machines) worldwide in one network alone (the MasterCard Network [13]). In the restaurant industry, it was reported that there are more than 690,000 restaurants in the United States alone [14].

Of course, not all hotspots will have a distinct network name. Typically, hotspots of the same WISP will use the same network name. However, in the market there may be a certain proportion of hotspots that are operated by small and independent operators. For example, owner of venues such as coffee shops or restaurants may want to operate their own hotspots. These independently operated hotspots tend to have their own distinct network names, and there can be millions of these hotspots globally.

A WISP can reduce the size of its roaming table by dividing it into sub-tables according to geography. For example, it can use one sub-table for each metropolitan area, state, or country. The finer the granularity of the geographical region, the smaller the table size. However, there is a trade-off. A traveling user visiting a new region has to pre-download the right table onto her MTs. The smaller the table, the more inconvenience this pre-download requirement will rest upon the users (e.g. the users need to pre-download more frequently; or they have a higher chance of forgetting to pre-download the right tables; etc).

There will also be a substantial management overhead to distribute and update these roaming tables on the users' MTs. To reduce manual attention, the WISP can build some intelligence into its client so that the table update is done in the background and is not sensitive to connection interruptions. Preferably the updates are streamed in small incremental data blocks, so that the update will not impact the performance of the client software even in low-bandwidth situations.

The storage requirement for these roaming tables may also be a problem. The problem is especially acute for handheld devices such as PDAs or smartphones, where non-volatile memory spaces are more limited. For example, if there are a million table entries in the table, and each entry on average takes up 10 bytes for the network name, then the table size will be at least 10 Mbytes.

Overall, so far the roaming-table solution has been working for a scale where only a few thousands of hotspots are involved. It is not clear whether this solution can be scaled up to handle situations where millions of hotspots are involved.

Addressing the scalability problem of using roaming tables, recently several solutions have been proposed by different groups. We will discuss these proposals in the rest of this section. There will also be a discussion on some other related work in Section 8.

4.2 Multiple SSIDs and Virtual AP

A roaming organization may convince their member WISPs to change their practice. Instead of broadcasting their WISP-specific network name, these WISPs will broadcast the network name of the organization. An example is illustrated in Figure 6. Being a member of Organization-Y, WISP-B and WISP-D would both configure their APs to broadcast the name "Y". They would also configure the MTs of their users to select an AP of the network name "Y". In this case, these users do not need to carry around the roaming tables discussed in the previous section. Instead, they only need to configure the SSID "Y" into their MTs, and these MTs will be able to get services in all hotspots under Organization-Y (B,C,D,P,Q,R,S,T).

However, while this approach works for WISP-A, WISP-B, and WISP-D, it may not work for WISP-C, unless the APs of WISP-C can announce multiple SSIDs. This is because WISP-C has roaming agreements with both Organizations X and Y.

We believe a good solution for discovery of roaming information should support a WISP like WISP-C. This will be beneficial to both the WISPs as well as the roaming organizations. For a WISP, it can get more revenues because more users will be able to use its network. For a roaming organization, it can reduce the risk of being locked out of a given venue.

However, the standard does not specify a standard way for a given AP to support the use of multiple SSIDs. Therefore, while there have been some products on the market that support multiple SSIDs, they were not implemented in the same way. These include Cisco 1100 Series, Symbol Mobius's Axon Wireless Switch, and the product from Col-

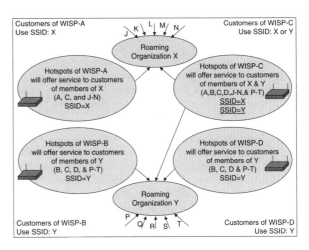

Figure 6: APs of WISP-C broadcast multiple SSIDs

ubris Networks. Without a standard, there is no guarantee of interoperability between equipment of different vendors.

To ensure inter-operability, in the IEEE 802.11 Working Group, there is a proposal on standardization of the support of multiple SSIDs. B. Aboba called his proposal "Virtual AP" [2]. In essence, his proposal is to put multiple "virtual APs" (VAPs) into one physical AP. Each VAP will have its own SSID, BSSID, and capability information, and each VAP will broadcast its own beacon. To the MT, it cannot even tell that the different beacons that it receives are from different distinct APs or are from different VAPs hosted on a single physical AP. Since each "virtual AP" has its own BSSID, once the client has decided on using the VAP, it will use the BSSID for in all data, control, and management frames communicating with the AP. As a result, both the AP and the client can, at the link level, filter out frames of BSSIDs that they do not recognize without first decrypting them.

The Virtual-AP proposal retains the important flexibility of allowing an AP operator to join more than one roaming organization. Going back to our roaming example, this means WISP-C can be members of both Organizations X and Y. Its AP will announce the fact by broadcasting both network names "X" and "Y," as illustrated in Figure 6. However, there are a number of drawbacks:

- Long effective inter-beacon interval: Since different beacons are broadcast in a round-robin fashion, if inter-beacon interval is T ms, and N virtual WISPs are supported for a given AP, then the effective inter-beacon interval for a particular VAP will be $N \times T$ ms. Practically, it is expected that a given AP cannot host more than 8 or 16 different VAPs [8].

- Bandwidth inefficiency: The beacon traffic competes the same air bandwidth with data traffic. So the more beacons an AP sends out, the less bandwidth will be left for data traffic. (For example, if $N = 3$, and we want to keep $N \times T = 60ms$, then we will have to set $T = 20ms$. A simulation study in [17] shows that when $T = 20ms$, 20% of an AP's capacity will be used for beacons alone.)

- No backward compatibility: Finally, this approach is not backward compatible with existing APs, since most of them do not support the broadcasting of multiple SSIDs. To adopt this scheme, the WISP or APO will have to upgrade their the hardware or firmware of their equipment. Most software clients running on the MT also do not understand multiple SSIDs and they will also need to be upgraded.

There is a possible variation to the VAP scheme described above. In this variation, the AP is configured with multiple VAPs, each for a different WISP, and each has its own SSID, BSSID, etc. However, the AP will not broadcast the SSID of these VAPs. Instead, it will only reply with a probe-response when an MT sends it a probe request embedded with the configured SSID of the VAP. While this variation can avoid the problem of bandwidth inefficiency described above, it has other drawbacks. Broadcasting of SSID can be a crucial means to attract users to use the AP — without it the user may not even be aware that a service is available. Now the MT has to constantly probe all APs within range. Also, some WISPs may insist to have their SSID broadcast.

4.3   Overloading EAP-Request/Id

In the EAP (Extensible Authentication Protocol) Working Group of the IETF, there is also a new proposal being considered [3]. There is not yet a suggested name for the proposed solution, so we are referring to it as "Overloading EAP-Request/Id".

EAP-Request/Id is a mandatory message sent from an 802.1x-compliant AP to a Wi-Fi client [16]. Its purpose is to query the identity of the client as a preparation step to complete the EAP-based authentication (RFC3748). The payload in this message is called "Type-Data," which is used to convey a displayable message for the client. The message can be terminated by a NULL character, but its length is not determined by the NULL character but by the "Length" field in the header. Incidentally EAP does not specify the use of the octets after the NULL.

Exploiting this flexibility, the proposal of "Overloading EAP-Request/Id" suggests to put into these octets the roaming information (Figure 7). It further defines a text-based representation scheme as an attribute called "NAIRealms" (for Realms of Network Access Identifier). In the attribute, the domain names of the WISPs that have roaming agreements with the operator of the AP are listed. Using our example of Figure 4, the attribute for the hotspot of WISP-C would be

`NAIRealms=RoamOrg-X.org;RoamOrg-Y.org`

Since EAP messages cannot be fragmented, each message is limited to the size of the network MTU (maximum transfer unit), which can be as small as 1020 octets. As a result, there is an implicit limit on the number of domain names (maximum size 255 octets) that can be listed [8].

This approach makes use of the 802.1x/EAP mechanism, so it is backward compatible with the existing APs if they support 802.1x (which is a requirement of WPA). However, it is not backward compatible with earlier non-802.1x APs.

The main disadvantage of this approach is that the MT has to associate, on the link layer, with the uncontrolled port of the AP first before EAP messages can be exchanged. That means the MT has to associate with all candidate APs one by one before it can acquire all the roaming information. The problem is especially acute when fast handoff between

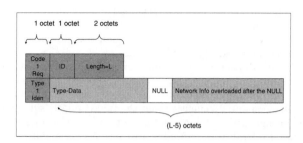

Figure 7: Overloading EAP-Request/Id

APs is desired, because the MT has to first give up its association with its current AP before it can discover the roaming information of a candidate AP. This will lead to large inter-AP handoff latency.

Another disadvantage of this solution is that there may be a contention problem. This will happen if the space after the first NULL is already used for some other non-standard purposes.

4.4   Brute-Force / Try-and-See

There is also a proposal that can be categorized as "Brute-Force" or "Try-and-See." In this proposal, an MT will ignore the SSIDs broadcast by the APs that are seen within the radio range. It simply tries to associate and then authenticate with these APs one by one. If it succeeds with one, then the very fact that it can authenticate confirms that the AP has a roaming agreement with the user's WISP. [2]

The advantage of this approach is that it is very simple, and it is fully compatible with all the existing APs (802.1x or even the earlier non-802.1x ones).

However, the disadvantage is that the MT has to first associate with a candidate AP, and thus disassociate with it current AP, before it can acquire the roaming information from the candidate. Furthermore, each attempt may take a few seconds to complete. This is because it may involve a few round trips of signaling all the way back to the home WISP. The MT cannot hide the latency from the user because it has to disassociate first with the current AP.

The user may have multiple subscriptions with different WISPs, each has a different set of credentials (username and password). In this case, the brute-force approach will have another problem. That is, the MTs does not know how to choose the right set of credentials since it does not know with whom it is try to authenticate with. It can only resort to trying these credentials one by one. This further increases the association latency.

4.5   Manual / Out-of-band Approach

Finally, there are manual solutions. For example, the operator of a hotspot can post a sign saying

This hotspot supports WISP-A, WISP-B, WISP-C, WISP-D. Please set your SSID to "McBurger"

The users are expected to manually configure the the SSID of their MTs when they roam to these partner networks.

---

[2]We ignore in this discussion the authentication of the AP to the MT.

The advantages of this approach are that it is very simple, and it is fully compatible with all the existing APs (802.1x and the earlier non-802.1x ones). The disadvantages are that it requires manual intervention from the users, and it will not provide automatic handoffs between APs of different WISPs.

5. OUR PROPOSAL: RIC

In this section, we are going to propose our own solution — *Roaming Information Code (RIC)*, which has two basic forms: *Bitmap* and *Concatenated Roaming Organization Numbers (RON)*. Furthermore, there are two possible options for a RIC to be transported from an AP to an MT — it can be put into the SSID space, or it can be put into a new (to be standardized) Information Element (IE). The key idea is that a space of, say, 32 octets (256 bits) can encode $2^{256} = 1.16 \times 10^{77}$ different patterns. If we choose our representation scheme carefully, we can encode the much needed roaming information into this space.

5.1 Roaming Information Code (RIC)

The RIC has a space of r octets. Without loss of generality, let us assume for now that $r = 32$. In the basic scheme, the 32-octet RIC space is divided into four parts: p octets for *Prefix*, f bits for *Flags*, v bits for *Version number*, and c bits for *Code*. $8p + f + v + c = 32 \times 8 = 256$. (See Figure 8)

- The *Prefix* is used to expand the space of RIC. Each prefix creates a new code space. We can assign the prefix geographically or non-geographically. For example, if we set $p = 2$, and assume we use only the ASCII characters [a-z][A-Z][0-9] (i.e. totally 62 different alphabets), then we can have $62 \times 62 = 3844$ different prefixes. Initially, we can use the well known ccTLD ("country code — Top Level Domain" used in the Internet domain name system) for each country. That is, "US" is assigned to the USA, and "UK" is assigned to Britain, etc. The other patterns are reserved for future uses.

- The *Flags* can be used for control information. For example, we can have a "check bit" to help differentiating a RIC from a legacy SSID. The idea is to define the most significant bit of the Flags octet to be the check bit. The bit is always set. This makes sure that the octet in every RIC is always a non-ASCII character. As such, a RIC will not be confused with a legacy SSID, which is unlikely to contain non-ASCII characters. [3] We can also have a bit to differentiate which of two forms of Code is used (see below). Unused bits will be reserved for future use. (E.g. $f = 8$).

- The *Version number* allows changing of interpretation as time goes by. (E.g. $v = 8$).

- The *Code* contains the actual coding of roaming information. This can further take two different forms. (E.g. $c = 224$).

---

[3] For example, without the check bit, a legacy SSID "USA11," or in hexadecimal "55 53 41 31 31 00H," can be confused as a RIC with the following interpretation: prefix is "US", bits 1 and 7 of flags are set (41H), version number is 49 (31H), and bits 2,3,7 of the bitmap are set (31H). With a check bit this confusion will not happen.

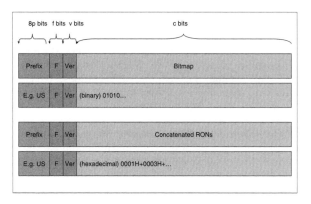

Figure 8: Basic scheme of RIC, which has two possible forms.

Under each form of the RIC, we also show an example (assuming WISP-1 and WISP-3 are supported). We show binary and hexadecimal number as indicated. The "+" sign represents concaternation.

Bitmap. In the first form, the Code is interpreted as a bitmap. Each bit in the map is assigned to a roaming organization (such as X and Y in Figure 4). A set bit means "this AP is a member of this organization"; a clear bit means "this AP is *not* a member of this organization". If $c = 224$, then at most 224 distinct roaming organizations can be represented per prefix.

Concatenated RONs. In the second form, each roaming organization is assigned a *Roaming Organization Number (RON)* in an a-octet name space. This will mean that there can be 2^{8a} different such organizations for each prefix. These RONs can be concatenated and fit into the space of Code. The RON of an organization is listed if the operator of the AP has a roaming agreement with the organization. The concatenation scheme means that the length of the Code part will limit how many RONs can be listed. In other words, at most n roaming organizations can be supported per AP, where $8a \times n \leq c$. Assuming that we set $a = 2$ and $c = 224$, then under each prefix there can be at most $2^{16} = 65536$ different RONs, and there can be at most $224/16 = 14$ RONs listed in Code.

The above is our proposed basic scheme for RIC. We believe the final exact format have to be defined by an industrial consortium (e.g. Wi-Fi Alliance) after discussions among various parties in the industry. There are some possible variations to the basic scheme. In this paper, due to limitation of space, we will have to omit the discussion about them.

As for the bit or RON assignment, the consortium may choose its own policies. For example, it may delegate the assignment to a country-level organization. It may assign bits or RONs on a first-come-first-serve basis or by auction. It may assign the bits or RONs on a lease basis, meaning that a roaming organizations will have to pay a "rent" to maintain an assignment. It may also use a combination of these policies.

Note that we expect that only roaming organizations, but not individual WISPs, [4] will need bit or RON assignments. We also expect that the number of distinct roaming organizations is much smaller than that of distinct WISPs — while we may have millions of WISPs globally, we expect to have only a handful of roaming organizations. We therefore do not expect the exemplary limits (224 bits per prefix in the bitmap case; 65,536 RONs per prefix in the case of concatenated RONs) will pose severe scalability problems or hinder the growth of the industry. Similarly, we also do not expect that a scalability problem will be imposed by the exemplary limit that at most 14 RONs can be concatenated in a RIC. This is because a given operator of a hotspot may not want to have agreements with too many different roaming organizations, because each of it will require a separate setup of billing and accounting channels. Note also that to be more "future- proof", we may relax r from 32 octets to, say, 64 octets.

We believe these expectations are practical. In fact they are also implicitly assumed in the other solutions proposed in Section 4. For example, practically there cannot be more than 8 or 16 virtual APs hosted on a given AP [8]. As another example, the number of roaming organizations that can be listed in the text-based NAIRealms attribute is limited by the network MTU (maximum transfer unit) [8].

5.2 Transportation of RIC

We propose two options for RIC to be transported from an AP to an MT.

RIC as SSID. In this case, the SSID field of beacon and probe response will not be carrying the traditional network name. Instead, it will be carrying RIC. This solution is fully backward compatible with the existing APs, both the 802.1x and the earlier non-802.1x ones. However, the operators of the APs can no longer keep their traditional network names. Furthermore, operators have to perform a one-time change to re-configure their SSIDs of these APs.

RIC as an IE. The AP can announce RIC as a new (to be standardized) information element (IE) of the 802.11 standard. This solution allows operators to retain their traditional network names and to announce them in SSIDs. However, it will require a technology standardization for the new IEs. Also, the hardware and/or firmware of the existing APs will have to be upgraded.

In both cases, the MT will need a new client software component that understands the RIC and makes the network-selection decision accordingly. Figure 9 illustrates the idea assuming RIC is transported as SSID.

5.3 Pros and Cons of RIC

There are many advantages of using RIC to represent roaming information. First of all, it is very scalable. It does not suffer from the scalability problem as in the solution of using roaming table. Using our exemplary values for the various parameters, there can be as many as 224 roaming organizations represented per prefix using the bitmap form. If the form of "concatenated RON" is used, then there can

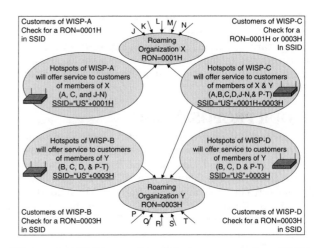

Figure 9: RIC is used and is transported as SSID

We assume the Code part of RIC uses concatenated RONs. To simplify the illustration, we do not show Flags and Version numbers in the RICs. In the RIC examples, "" denotes ASCII strings, + denotes concatenation, a numeric number with the H suffix is a hexadecimal number. We assume that the roaming organizations X and Y are assigned RON of 0001H and 0003H respectively.

be as many as 65,536 roaming organizations represented per prefix, and each AP can have roaming agreements with as many as 14 different roaming organizations.

Second, the MT does not have to first associate with an AP before the MT can acquire the roaming information (as in the cases of "Overloading EAP-Req/Id" and "Brute Force"). As a result, the RIC solution does not hinder the implementation of fast handoffs.

Third, the representation scheme is very compact. Using only 32 octets, RIC can encode a lot of information. It is far more compact then other text-based schemes, such as the NAIRealms attribute proposed as part of the solution "Overloading EAP-Request/Id."

Fourth, the RIC solution is very efficient. There is no roaming tables to be carried around by the MTs (as in the solution of using roaming table). There is no additional beacons or additional probe-requests/responses to be needed (as in solution of "Virtual AP"). There is no additional level-2 exchanges between the MT and the AP (as in the solution of "Overloading EAP-Request/Id"). There is no need for human intervention (as in the manual solutions).

On the other hand, the RIC solution may have some shortcomings. It will require a consensus from the industry to adopt the solution. It will also require an industry consortium to manage the bit and/or RON space. However, since the solution is very simple and very easy to understand, we would expect that the process is relatively easier than that of the other solutions that require new technologies.

5.4 Test implementation

To confirm that there are no hidden problems in the 802.11 standard, driver, or card firmware that prevent us from putting non-ASCII characters into the SSID space, we im-

---

[4]perhaps except some very large nationwide or global WISPs

Solution	Scalability w.r.t. Number of WISPs	Backward Compatibility	Acquires Information without Association	Fast Handoff Possible
Roaming Table	- Poor	+ All APs	+ Yes	+ Yes
Virtual APs	+ Good	- No	+ Yes	+ Yes
EAP-Req/Id	+ Good	- only 802.1x APs	- No	- No
Brute force	+ Good	+ All APs	- No	- No
Manual	+ Good	+ All APs	+ Yes	- No
RIC as SSID	+ Good	+ All APs	+ Yes	+ Yes
RIC as new IE	+ Good	- No	+ Yes	+ Yes

Figure 10: Comparing the various solutions for the discovery of roaming information. To aid visualization of the pros and cons, we precede each characterization with a "+" for pros and a "-" for cons.

Solution	Other Pros and Cons
Roaming Table	- Large storage requirement - Overhead of distribution and maintenance of roaming tables on MTs
Virtual AP	+ Partition of traffic at the link layer according to BSSID - Long inter-beacon intervals - Bandwidth inefficiency (if extra beacons are used) - Presence of service is not announced if WISP-specific SSIDs are not broadcast.
EAP-Req/Id	- MT has to associate with APs one by one to acquire information - May have contention problems
Brute force	- MT may have to try credentials one by one
Manual	- Require user's manual intervention - No automatic inter-WISP handoffs
RIC as SSID	+ Compact representation scheme + Efficient - Operators of AP can no longer choose their own network names - Requires industry consensus - Requires a management body to handle bit or RON assignments - Operators of AP have to perform a one-time change from network name to RIC
RIC as new IE	+ Compact representation scheme + Efficient - Requires industry consensus - Requires a management body to handle bit or RON assignments

Figure 11: The other pros and cons of the various solutions for the discovery of roaming information. To aid visualization of the pros and cons, we precede each characterization with a "+" for pros and a "-" for cons.

plemented a RIC demo on our test-bed. The AP was a Linux system equipped with a Prism2-based Wi-Fi card and the HostAP software. The client was a Linux laptop with an Orinoco card. We used the `orinoco_cs` driver (version 0.15rc) [1], which supports the AP scanning mode. We modified the user-level utilities `iwconfig` by adding a new command option `iwconfig <interface> ric PP FF VV BB`. The option sets a RIC into the SSID space (where PP, FF, VV, and BB stands for prefix, flags, version, and bitmap respec-

tively). We also modified the user-level utilities `iwlist` so that it can interpret a RIC when it sees one. The test implementation worked as we expected. There were no changes needed in the driver or card firmware.

5.5 Comparing the various solutions

Figures 10 and 11 summarize the various proposals that we have discussed. From the table, we can see that each solution has their own pros and cons. Since different parties in the industry may weight these pros and cons differently, we advocate a discussion in the community to evaluate the various proposals. Overall, RIC seems to be the most promising solution; however, Virtual AP and Roaming Table may also be acceptable. The goal of this paper is to highlight the importance of the issues, to contrast these different proposals, and to facilitate a discussion within the community so that a good solution can be agreed upon.

6. DISCOVERY OF SECURITY POLICIES

So far our discussion has been focused on the discovery of roaming information, which is the most critical information needed by an MT for a decision on network selection. However, as the industry continue to mature, there may be other needs to be addressed. In this section, we will discuss how a given AP can allow different WISPs to have their own specific security policy, and how an MT can discover this information. In the next section, we will discuss how an AP can announce price and workload information.

In Section 4.2, we discussed the mechanism of *Virtual Access Point (VAP)*. While it may not be the best solution for the discovery of roaming information, it has merits for other applications. For example, it allows partitioning of traffic at the link layer. It also allows different WISPs hosted on a given AP to choose their own security schemes.

As discussed in Section 2, there are different security schemes for 802.11: UAM, WEP, WPA, and 802.11i. It is very likely that these different schemes may co-exist in the market for a long period of time. And it is likely that different WISPs hosted on a given AP may want to mandate their own security schemes.

Without VAP support, all WISPs supported under a given AP have to use a common security scheme. This is because the traditional AP can only announce one SSID and one Capability or RSN IE describing the mandated security scheme (see Section 2).

With VAP support, this restriction can be relaxed. We propose that VAP can be used in combination with RIC to achieve efficiency and flexibility. We called the combined scheme *RIC-VAP*. In the following discussion, we assume the following model of VAP support: There is a single SSID per beacon but the AP can reply Probe Response for different VAPs, each with a distinct BSSID and Capability or RSN IE. The procedures of RIC-VAP for the discovery of roaming information and security policy are as follow:

1. Each WISP will have a distinct SSID, which is also well known to the MTs. On a VAP-supporting AP, the WISP will be assigned a VAP with its distinct SSID, BSSID, and security scheme (described in the Capability Info IE or the RSN IE). However, the AP will *not* broadcast beacons containing these WISP-specific information. Instead, it will just broadcast a single generic beacon containing roaming information in RIC, which can be transported as SSID or as a new IE as discussed in Section 5.

2. Through the RIC, an MT can determine whether the AP supports the roaming organization of the user's WISP.

3. If the answer is positive, then the MT may decide to select the AP network. It will then send a Probe Request with a SSID that is specific to the WISP.

4. Upon receiving the Probe Request addressed to itself, the VAP on the AP would reply with a Probe Response containing its own SSID, BSSID, and security scheme described in the Capability Info or RSN IE.

5. Upon receiving the probe-response, the MT will proceed to associate/authenticate with the AP, using the acquired BSSID and the suggested security scheme. All subsequent signaling and data traffic between the MT and the AP will use the same BSSID.

Comparing to the pure VAP scheme, the RIC-VAP scheme has the advantage that the AP does not need to broadcast beacons of different VAPs in a cycle, yet the presence of the WISP is announced through RIC. Therefore, this scheme can avoid the long inter-beacon intervals and the bandwidth loss due to the broadcast of extra beacons.

Comparing to the pure RIC scheme, the RIC-VAP scheme has the advantage that different WISPs can now choose their own security schemes. Furthermore, there is link-layer partitioning of traffic.

7. DISCOVERY OF OTHER SERVICE INFORMATION

Going forward, people may want their decisions of network selection be based on more detailed dynamic information. For example, they may want to choose AP networks based on the current price. In this case, we will need a mechanism for an MT to discover dynamically the price information from an AP. As another example, there has been some discussion suggesting that an MT should select an AP based on the workload of the AP [10, 4]. Again, we will need a dynamic discovery mechanism here.

Unlike some other proposals (e.g. the CARD protocol that we will briefly discuss in Section 8), our proposal for the discovery of this information is strictly done in layer-2 (link layer). We think this is more appropriate because the information is needed for a link-layer decision (network selection).

Our proposal is to extend the existing 802.11 standard with new Information Elements (IEs). Like other existing IEs, these IEs will be used in beacons and probe-responses. We propose two new IEs here, each with a suggested format, which can serve as a baseline for adoption in the 802.11 working group.

These new IEs can be used with or without the support of virtual AP (Section 4.2). When used together with the support of VAP, they allow individual WISP to announce its own information. For example, each WISP can set their own dynamic price. Even if VAP is not supported, they may still be useful because the operator of the AP may want to announce a generic but dynamic information (e.g. surcharge information).

7.1 Compact Code for Price Information

For price information, we propose a new IE called Compact Code for Price Information (CCPI). Like in other 802.11 IEs, the first octet of this IE is for *ElementID*, and the second octet is for *Length*. Following these there are five fields:

- V bits for *Version*. (E.g. $V = 4$)

- F bits for *Flags*. (E.g. $F = 4$) We can have a flag indicating whether the advertised price is a surcharge or a total charge.

- N bits for *Numeric*. (E.g. $N = 32$) We may consider using the IEEE Single Precision Floating Point Number.

- U bits for *Unit*. (E.g. $U = 4$) Totally 2^U different units can be represented. For example, 0 for "per minute", 1 for "per hour", 2 for "per day", etc.

- C bits for *Currency*. (E.g. $C = 12$) Totally 2^C different currencies can be represented. For example, 0 for US Dollar, 1 for British Pound, 2 for Euro, etc

This element is of length $2 + L$ octets, where $L = \lceil (V + F + N + U + C)/8 \rceil$. (In our example, $L = 7$).

Compared to the other text-based representations of price information, this representation is very compact yet very expressive. We believe an encoded scheme like the one we propose is more appropriate for use in the link layer.

7.2 Compact Code for AP Workload

For workload information, we propose a new IE called Compact Code for AP Workload (CCAPWL). Like in other 802.11 IEs, the first octet of the IE is for *ElementID*, and the second octet is for *Length*. Following these there eight fields:

- V bits for *Version*. (E.g. $V = 4$)

- F bits for *Flags*. (E.g. $F = 4$)

- 3 fields for parameters related to *Over-the-air Workload*, each has U bits. (E.g. $U = 8$). We propose to have three parameters so that the AP can announce three running averages (e.g. for the past 1, 5, and 15 minutes).

- 3 fields for parameters related to *Backhaul Workload*, each has W bits. (E.g. $W = 8$). We propose to have three parameters so that the AP can announce three running averages (e.g. for the past 1, 5, and 15 minutes).

This element is of length $2 + L$ octets, where $L = \lceil (V + F + 3U + 3W)/8 \rceil$.

8. RELATED WORK

There have been some discussions on the issue of network selection and discovery of service information in various standard bodies such as IETF, IEEE, and 3GPP2.

The Internet Draft [8] summarizes the discussion held in the EAP (Extensible Authentication Protocol) working group of the IETF. It discusses the problem as well as some solutions but it does not propose its own solution.

Within the same working group, there is an individually submitted Internet Draft [3] that proposes to overload the EAP-Request/Id message for carrying roaming information. The main disadvantage of this approach is that it will require the MT to associate with the AP first before acquiring information. We discussed this solution in Section 4.3.

In another IETF working group — Seamoby (Context Transfer, Handoff Candidate Discovery, and Dormant Mode Host Alerting), there is a proposal called CARD, or Candidate Access Router Discovery [11]. The goal is to allow an MT to perform seamless handoff from one AP to another. CARD provides an IP-level protocol for the MT to discover,

from the currently attached AP, the identities and capabilities of another AP (which is called CAR, or candidate access router). The result is that the MT can know some information about the CAR before the MT initiates attachment to the CAR. In this sense, CARD tries to achieve a similar goal as we discussed in Sections 5, 6, and 7. However, CARD is an IP-level protocol, meaning that a lower layer (link-layer) decision will have to be depending on the operation of a higher layer (IP-layer) protocol. Furthermore, CARD's operation depends on the availability of a current AP, with whom the MT can query. However, when an MT first enters a hotspot, it is not attached to any current AP. CARD does not appear to provide a complete solution for network selection.

The IETF working group PANA (Protocol for carrying Authentication for Network Access) is designing an IP-level authentication protocol that is agnostic to the link layer. In the protocol, there is a mechanism for an AP to advertise "ISP" information in an "ISP-Information AVP (Attribute Value Pair)". We think that information needed for a link-layer decision should be acquired at the link-layer but not at the IP layer.

In IEEE, there has already been a discussion on the standardization of the broadcasting of multiple SSIDs for a given AP. The proposal is called "Virtual AP." As we discussed in Section 4.2, the use of VAP requires technology standardization. Furthermore, the use of multiple beacons may increase the effective inter-beacon intervals and may reduce the effective bandwidths of the airlinks.

In the specification prepared by the task group K of the IEEE 802.11 working group (Radio Resource Management), there are mechanisms for a client to report layer-1 and layer-2 statistics to an AP, and for a client to request a "site report" from the AP for these collected statistics. This may address the issue related to AP workload.

The task group AB of the IEEE 802.1 working group (Station and MAC Connectivity Discovery, also known as LLDP, or Link-Layer Discovery Protocol) is defining a protocol for discovering the physical topology and connection end-point information from adjacent devices in 802 local-area (LANs) and metropolitan-area (MANs) networks. This work is for all IEEE LANs and MANs, wired or wireless. We believe wireless LANs (802.11) has its special needs for discovery of service information because of the special nature of wireless link (e.g. the notions of association and disassociation).

In 3GPP2, there has been a proposal to use SSID as well as the (non-yet standardized) virtual AP mechanism for network selection [12]. However, in the proposal there is no discussion about the drawbacks that we presented in Section 4.2.

9. CONCLUSIONS

In this paper, we highlighted the importance of the discovery of service information, and its relevance to the issue of network selection. The most critical service information is roaming information; while other information such as security policies, price, and AP workload may also be needed.

For roaming information, we discussed the various existing proposals for its discovery: Roaming Table, Virtual AP, Overloading EAP-Request/Id, Brute Force, and Manual. And then we proposed our own solution — Roaming Information Code (RIC), which can be transported as SSID or a new (to be standardized) Information Element of the

IEEE 802.11 standard. We contrasted the pros and cons of the various solutions. We favor our own solution because it is scalable and is fully backward compatible with existing APs (if RIC is transported as SSID), and it does not hinder fast handoffs.

We then proposed another two schemes for the other service information. We proposed a RIC-VAP scheme that combines the best features of RIC and VAP to allow different WISPs to have their own specific security policies on a given AP. We also proposed to add two new Information Elements in the 802.11 standard for an AP to announce price information and AP workload information.

Overall, this paper is about a basic principle: an MT should have some basic information about an AP and its network before the MT can decide whether or not to attach to the AP. All our proposals operate under this principle. We hope this paper will help foster a discussion in the community on the important issue of network selection in public WLAN hotspots.

10. ACKNOWLEDGMENTS

We would like to thank Peter J. McCann and Girish P. Chandranmenon for their insightful comments that helped us revise the paper to a much better shape. We would also like to thank the anonymous reviewers for their various suggestions.

11. REFERENCES

[1] The Linux Orinoco Driver. http://www.nongnu.org/orinoco, 2004.

[2] Bernard Aboba. Virtual access points. IEEE Contribution 11-03-154r1, http://www.drizzle.com/~aboba/IEEE/ 11-03-154r1-I-Virtual-Access-Points.doc, May 2003.

[3] Farid Adrangi. Mediating Network Discovery and Selection. draft-adrangi-eap-network-discovery-and-selection-01.txt, Individually submitted Internet Draft, February 2004.

[4] Victor Aleo. Load Distribution in IEEE 802.11 Cells. Master of Science Thesis, KTH Royal Institute of Technology, Stockholm, Sweden, March 2003.

[5] Wi-Fi Alliance. Wi-Fi Protected Access. http://www.wifialliance.com/OpenSection/ protected_access.asp, 2004.

[6] B. Anton, B. Bullock, and J. Short. Best Current Practices for Wireless Internet Service Provider (WISP) Roaming. Version 1.0, Wi-Fi Alliance, February 2002.

[7] W.A. Arbaugh, N. Shankar, Y.C.J Wan, and Z. Kan. Your 802.11 wireless network has no clothes. *IEEE Wireless Communications*, 9(6), December 2002.

[8] J. Arkko and B. Aboba. Network Discovery and Selection Problem. draft-ietf-eap-netsel-problem-00, Internet Draft of the IETF EAP Working Group, January 2004.

[9] Nikita Borisov, Ian Goldberg, and David Wagner. Intercepting Mobile Communications: The Insecurity of 802.11. In *Proceedings of the 7th Annual International Conference on Mobile Computing and Networking*, Rome, Italy, July 2001.

[10] Glenn Judd and Peter Steenkiste. Fixing 802.11 access point selection. A poster in the SIGCOMM Poster Session, http://www-2.cs.cmu.edu/ glennj/scp/ FixingAPSelection.pdf, 2002.

[11] Marco Liebsch, Ajoy Singh, Hemant Chaskar, Daichi Funato, and Eunsoo Shim. Candidate Access Router Discovery. draft-ietf-seamoby-card-protocol-06.txt, Internet Draft of the IETF Seamoby Working Group, December 2003.

[12] Serge Manning. WLAN Network Selection. 3GPP2 contribution 3GPP2-X31-20040419-xxx, April 2004.

[13] MasterCard International, Inc. ATM Locator – Frequently Asked Questions. http://www.mastercard.com/atmlocator/ index.jsp?page=faqs, 2004.

[14] National Restaurant Association. Eating and Drinking Place Statistics, By States. http://www.restaurant.org/research/ state_stats.cfm#establishments, 2004.

[15] Institute of Electrical and Electronics Engineers. Part 11: Wireless LAN Medium Access Control (MAC) and Physical Layer (PHY) Specifications. ANSI/IEEE Std 802.11, 1999 Edition, 1999.

[16] Institute of Electrical and Electronics Engineers. IEEE Standard for Local and metropolitan area networks: Port-Based Network Access Control. IEEE Std 802.1X-2001, 2001.

[17] Héctor Velayos and Gunnar Karlsson. Techniques to reduce ieee 802.11b mac layer handover time. KTH Technical Report TRITA-IMIT-LCN R 03:02, KTH, Royal Institute of Technology, Stockholm, Sweden, April 2003.

[18] Wi-Fi Alliance. Wi-Fi Overview. http://www.wi-fi.org/OpenSection/ why_Wi-Fi.asp?TID=2, 2004.

A Scalable Framework for Wireless Network Monitoring

Camden C. Ho, Krishna N. Ramachandran, Kevin C. Almeroth, Elizabeth M. Belding-Royer

Department of Computer Science
University of California, Santa Barbara
{camdenho, krishna, almeroth, ebelding}@cs.ucsb.edu

ABSTRACT

The advent of small form-factor devices, falling hardware prices, and the promise of untethered communication is driving the prolific deployment of wireless networks. The monitoring of such networks is crucial for their robust operation. To this end, this paper presents VISUM, a scalable framework for wireless network monitoring. VISUM relies on a distributed set of agents within the network to monitor network devices and store the collected information at data repositories. VISUM's key features are its extensibility for new functionality, and its seamless support for new devices and agents in the monitoring framework. These features enable network operators to deploy, maintain, and upgrade VISUM with little effort. VISUM can also visualize collected data in the form of interactive network topology maps as well as real-time statistical graphs and reports. These visualizations provide an intuitive, up-to-date, and useful overview of a wireless network. We have implemented VISUM and used it to monitor a wireless network deployment at UC-Santa Barbara. In this paper, we describe the architecture of VISUM and report on the performance of the monitored network using information collected by VISUM.

Categories and Subject Descriptors

C.2.3 [**Computer-communication Networks**]: Network Operations—*network monitoring, network management*

General Terms

Management, Design, Measurement

1. INTRODUCTION

The marketplace is witnessing an explosive growth in wireless technology. With the advent of small form-factor devices, falling hardware prices, and the promise of untethered communication, wireless networks are undergoing prolific deployment in private homes, corporate offices, whole communities, and even entire cities.

For the robust operation of wireless networks, it is crucial that monitoring complements increasing deployment. Monitoring offers several benefits to network operators, system designers, and researchers. Monitoring can provide network operators with valuable insight into the state of the network, which can in turn increase understanding of the network's topology and usage. It can also enable operators to perform critical tasks, such as site surveying, billing/accounting, and fault detection/isolation, all necessary for the robust operation of the network. With monitoring, operators can check for compliance of system implementations with set standards. Compliance checks are important because wireless networks are typically formed by users who carry devices with heterogeneous hardware and software supplied by different vendors. System designers and researchers can use monitoring to improve protocols and systems through the analysis of collected network state. Furthermore, this state can help designers to develop realistic data traffic [4], user mobility [7], and wireless propagation models [7]. Network simulators, such as NS-2 [8] and GloMoSim [21], can then apply these models to simulate real-world network behavior more accurately [5].

The monitoring of wireless networks, however, is challenging. This is because of the rapid pace of development of wireless technologies and the short time-to-market of the developed products. The result is that wireless hardware vendors implement proprietary solutions with little standardization. To monitor networks with proprietary solutions, tools are required that are "tailored" specifically for such solutions. Consequently, operating and maintaining a set of such tools becomes cumbersome and scales poorly with increasing network size. Furthermore, the tools typically use proprietary information formats for representing collected data. This makes the correlation of monitoring information, represented in various formats, for network analysis particularly difficult.

For the monitoring of wireless networks, a framework is required that enables network operators to monitor heterogeneous devices in a manner that is generic across all devices supplied by different vendors. Moreover, this framework should scale well to cope with increasing network size yet require minimal maintenance by network operators. Our goal is to address the need for such a framework with *VISUM*. VISUM is based on a distributed architecture for monitoring wireless networks. VISUM delegates the monitoring functionality to a distributed set of agents that monitor devices and send collected information to monitoring repositories. Its generic architecture allows the seamless integration of new devices and agents with minimal configuration. It does this by using a novel XML (eXtensible Markup Language) based framework to abstract device idiosyncrasies. In addition to the collection and storage of monitoring information, VISUM processes the stored information to create a number of interactive and graphical real-time representations of the network.

We have implemented VISUM using Java. As a result, our implementation is easily portable across various operating systems. To demonstrate its utility, we have used VISUM to monitor a wireless local area network (WLAN). This WLAN, consisting of sixteen access points, is deployed in a typical office building on the University of California, Santa Barbara campus. The information gathered using VISUM has given us valuable insight into the performance of the network.

The rest of this paper is organized as follows. In Section 2, we review work related to network monitoring. Section 3 describes the various challenges in meeting our goal of developing a generic framework for wireless network monitoring. In Section 4, we provide a detailed description of the design and the architecture of VISUM. Section 5 describes our VISUM implementation. Section 6 presents observations from our analysis of the WLAN deployment at UC Santa Barbara based on information collected by VISUM. Finally, we conclude the paper in Section 7.

2. RELATED WORK

A wide array of monitoring tools are available for wired networks. Early tools developed are *traceroute* and *ping*. Visualization extensions for these tools such as LACHESIS[15] and GTrace[13] synthesize additional information, i.e. historical and geographical data, to provide intuitive representations of collected information. Other recent tools use standardized management protocols such as the Simple Network Management Protocol (SNMP) [2] and syslog [10] to achieve sophisticated monitoring requirements. These tools come in two main flavors: tools that rely on information from within the network, such as information collected from network routers (e.g. Border Gateway Protocol state); and tools that rely on end-to-end data collection to monitor the state of the network. Examples of the former are Rocketfuel [16] and MANTRA[14]. Examples of the latter are ScriptRoute [17] and King [6]. Such tools have led to several studies that give valuable insight into the performance of deployed protocols and networks [20, 11, 12].

In the area of monitoring single-hop wireless networks, there is a general lack of tools that are easily accessible to the community. Some proprietary tools are supplied by access points vendors such as Cisco, Netgear, and Lucent. These tools are typically installed on the device itself and allow information to be accessed via SNMP or the Hypertext Transfer Protocol (HTTP). Their effectiveness, however, is generally limited by insufficient documentation and the proprietary nature of such tools. Nevertheless, numerous studies have analyzed the performance of such networks using these tools [18, 19, 3, 9, 1]. There have been efforts to study the usage and mobility patterns of wireless networks in several environments, namely university campuses [3, 9], metropolitan areas [18], and public areas [1]. Each study monitored its specific wireless network environment for a predetermined period using a collection of some of the aforementioned tools.

3. CHALLENGES

Broadly, our goal is to develop a generic framework for monitoring single-hop wireless networks. In this section we discuss some of the challenges in achieving our goal. These challenges are as follows:

- **Network Size**: Because of the rapid pace of WLAN deployments and their universal appeal, it is not uncommon for networks to consist of hundreds of access points. As examples of networks of such scale, companies such as Boingo Wireless and Wayport have deployed networks that span whole communities and even entire cities. Networks of such scale can consist of devices supplied by different hardware vendors. Because of device idiosyncrasies, monitoring such networks is challenging. Furthermore, several of the solutions typically rely on a centralized infrastructure for monitoring. This can make the monitoring techniques scale poorly.

- **Device Integration**: As networks grow in size, it is crucial that new devices are seamlessly integrated into the monitoring solution. This can be difficult because of device idiosyncrasies in different vendor products and even across various products released by a single vendor. To overcome these differences, network operators may be required to manually configure and maintain monitoring tools.

- **Information Retrieval**: Information collected from heterogeneous devices arrive in inconsistent proprietary formats, making retrieval and utilization of the accumulated information challenging. The architecture should be general enough to make the collected information easily accessible. Higher level applications should not require detailed knowledge of network components for data mining and analysis. Therefore, the method of storing accumulated information must be considered carefully, keeping in mind the need for a robust and scalable structure that allows efficient data retrieval.

- **Extensibility:** Because of the rapid pace of development in wireless networking technology, it is difficult for a monitoring solution to remain useful for monitoring newly developed network devices. It is critical for a wireless network monitoring system to be easily extensible for the purposes of collecting information from newly developed devices. Similarly, the system should provide the necessary facilities to retrieve new information that is deemed important to WLAN performance.

In the next section, we describe the design of VISUM to overcome the challenges described above.

4. VISUM DESIGN

VISUM is based on a distributed architecture for monitoring large scale wireless networks. It delegates the monitoring task to a set of agents distributed throughout the network. These agents collect monitoring information from network devices using SNMP and store the collected information at a centralized repository. In this paper, our discussion of the VISUM architecture assumes that VISUM uses a centralized repository. However, it is easily extensible to support a

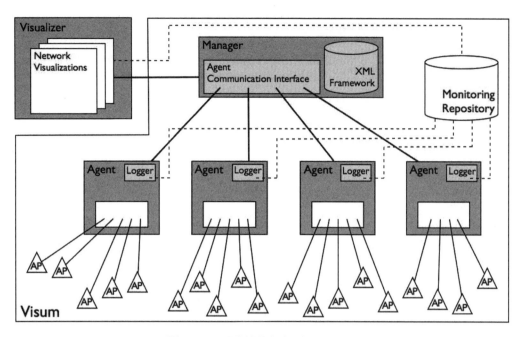

Figure 1: VISUM Architecture.

distributed set of repositories. VISUM is designed such that agents can be added seamlessly to the monitoring framework. This makes VISUM easily scale to large networks. Moreover, VISUM supports the seamless integration of new devices into its monitoring framework. Seamless integration of agents and devices is achieved using a novel XML-based framework that abstracts device idiosyncrasies.

Figure 1 illustrates a conceptual representation of the VISUM framework. The figure shows three main components in the framework. The *Manager* coordinates the operation of the distributed set of *Agents*. It also serves as a central location for the configuration and the maintenance of Agents. The *XML framework* is used to support the seamless integration of new agents and devices.

The remainder of this section describes the aforementioned components in more detail. The XML framework forms the foundation of our generic architecture. It is described first, followed by a detailed description of the remaining modules.

4.1 XML Framework

To enable VISUM to be easily extensible to a heterogeneous set of devices, we use an XML-based framework for abstracting device idiosyncrasies in device-specific XML profiles. These profiles map high-level monitoring information or *identifiers* that need to be retrieved to device-specific SNMP Object Identifiers (OIDs). By abstracting vendor-specific OIDs to vendor-independent identifiers, VISUM decouples the information retrieval process from the representation of information and its subsequent analysis. The XML profiles are organized according to generality in a hierarchical structure; profiles at the root level are more general, and profiles at lower levels becomes increasingly device-specific. Using this hierarchy structure, VISUM aggregates replicated OID mappings and enables partial data collection from network devices without specific XML profile definitions. A sample hierarchy is illustrated in Figure 2a. As examples

of profiles stored in this XML hierarchy, Figures 2b and 2c show common- and device-specific XML definitions. The OID mappings defined in XML profiles are organized according to device and information type. Each mapping consists of a single OID enclosed with its corresponding high-level identifier.

4.2 Agent-Manager Interaction

The Manager is used for the configuration and maintenance of the distributed set of agents. The only initial configuration needed when an agent is first deployed is the address of its Manager. Once the agent is installed and the Manager is aware of its presence, the Manager delegates a subset of network devices that need to be monitored. In this way, the Manager enables seamless integration of additional Agents without interrupting normal operation. The Manager also maintains the XML framework described above to provide device-specific OIDs for Agents to use for data retrieval. The Manager is responsible for notifying its Agents when changes in the XML framework occur.

4.3 Agent

Agents do the actual collection of monitoring information from network devices. Figure 3 shows the conceptual architecture of an Agent. It consists of three modules to collect monitoring information: the *InfoGather*, *NodeInfo*, and *Query Scheduler*. An additional *Logger* module is used for storing the collected information at the repository. Each Agent module is briefly described below. *InfoGather:* The InfoGather module retrieves monitoring information from network devices. The data retrieval process is illustrated in Figure 4. Information is gathered in the following manner: The InfoGather module uses the device description OID (system.sysDescr) to retrieve the description of the device it is querying (Steps 1 and 2). The device description OID is a standardized OID used by all device vendors to identify their devices. The InfoGather then uses the resulting device

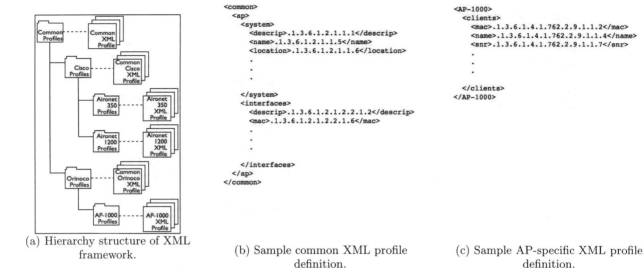

(a) Hierarchy structure of XML framework.

(b) Sample common XML profile definition.

(c) Sample AP-specific XML profile definition.

Figure 2: **VISUM XML Framework.**

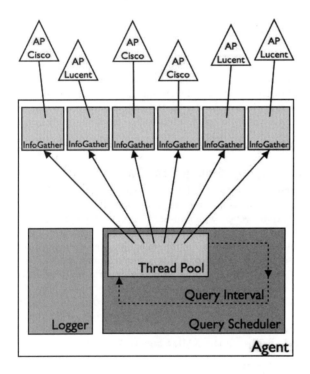

Figure 3: **The Agent Architecture.**

This configuration scheme has the advantage that Agents only require knowledge about each device's IP address and SNMP community string to configure itself to monitor the assigned devices. Furthermore, if the configuration of the network devices is modified, the Manager notifies the Agent of the change and the Agent adapts by retrieving the description of the new device and repeating the configuration process. *Query Scheduler:* The Query Scheduler is responsi-

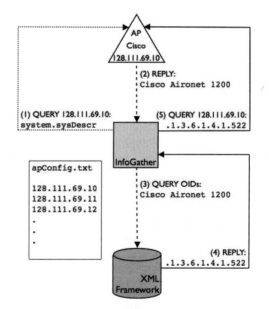

Figure 4: **VISUM XML profile-based InfoGather configuration.**

description to identify the appropriate set of XML profiles that contain the OIDs for the specific device. If the device-specific OIDs are not in the local OID lookup cache, the InfoGather module requests the device-specific OIDs from the Manager (Step 3). The manager retrieves the OIDs by traversing the profile hierarchy, first by the vendor type, then by specific model number. If a profile for the particular AP has been defined, the corresponding OIDs are retrieved and sent to the requesting Agent (Step 4). The Agent, in turn, stores the OIDs in a lookup cache to be used by the InfoGather module to monitor the AP (Step 5).

ble for the scheduling of the InfoGather modules to retrieve information from network devices. To capture the mobile nature of a wireless network, it is necessary for the Query Scheduler to schedule queries to all the monitored devices in the network both concurrently and with relatively high frequency. The scheduler maintains a pool of threads used to collect data from devices. In this way, the interval at which data is retrieved from the monitored devices is adjustable. The frequency is configured at the Manager by the network operator. There is a tradeoff between using high frequency querying, and the resulting increase in processing load on the device themselves and the resulting increase in network traffic.

NodeInfo: The data successfully retrieved by an InfoGather module is passed to the Logger module in the form of a NodeInfo object. NodeInfo is a generic representation of the data retrieved from the device. The NodeInfo object uses data types that can represent the values of various counters and gauges that are monitored in the wireless network devices. The Infogather module does the necessary type conversions so that data retrieved from the device can be represented using NodeInfo data types.

Logger: The Logger module's primary purpose is to process collected information before storing it. The Logger module exports a standard interface that can be used by the Info-Gather modules. The Logger module decouples the storage of data from the interface used to pass NodeInfo objects to the Logger module. Because of this, specific Logger modules can be used for storing the monitoring information in different storage formats including flat files and databases. Logger modules can also implement custom interfaces between VISUM and external applications or visualization tools.

5. IMPLEMENTATION

This section describes our implementation of VISUM. Our goal is to demonstrate the feasibility of developing a system based on the framework we propose. Our implementation of VISUM was developed completely in Java. The goal was to make Visum portable across various platforms. We have also leveraged Java's database support to implement our LoggerDB module. Descriptions of the implemented features are organized in the three phases: data collection, logging and presentation.

5.1 Data Collection Stage

Although we implemented an InfoGather module that utilizes the SNMP protocol to retrieve information from the APs in our test network, SNMP support in APs is far from absolute. If SNMP support is not available, additional components of the InfoGather module can be "plugged in" to monitor an AP that requires a different means of data collection. The NodeInfo modules are implemented as Java objects that encapsulate queried data. Each object contains specific information about a particular wireless network node. For example, a NodeInfo object representing an AP has information about the name, location, description and uptime. Similarly, a NodeInfo object describing a WLAN mobile host contains information about its MAC address, bytes sent and received, and average signal-to-noise ratio measured from the AP to the host.

5.2 Data Logging Stage

Our implementation has two Logger modules. The first module is for logging collected data in flat files. The second is for logging the collected data into a database. Although storing logged data in flat files is simple and efficient with regards to the data logging process, it limits the potential for

the system to be distributed. In this case, log files would be generated locally and would have to be later merged if each monitor is responsible for a subset of the entire network. The database Logger module is called LogDB. It supports remote database connections through the use of the Java Database Connectivity (JDBC) interface. Logging collected data into a robust database is a way to ensure that data remains intact, organized and always available for retrieval. Specifically, by storing monitoring information in a scalable and highly efficient database, VISUM can monitor networks for an extended period of time (possibly continuously given the reduced cost of storage) as well as provide up-to-date statistical graphs and reports efficiently.

5.3 Data Presentation Stage

The VisumGUI module handles the visualization of monitored data. It allows users to control VISUM system settings and interact with the real time visualizations implemented. The VisumGUI module leverages Java's graphical functionality to provide several options for the visualization of monitored data. Figure 5 shows screen shots of the visualization modules we implemented.

The real-time network topology visualization shown in Figure 5a represents a WLAN topology using a tree hierarchy. The root of the tree corresponds to the entire WLAN network, while the children of the root node represent each of the AP's actively monitored by VISUM. By expanding the AP nodes, the mobile hosts associated with the AP can be seen. The network topology tree is updated in real time. A user is able to view the details of a particular node by clicking and selecting the node of interest. Another network topology view maps the APs of a network together with its associated nodes in a graphical representation. Figure 5b is a snapshot of this visualization. The wireless links between an AP and its associated nodes are represented using various thicknesses to represent the quality of the links. The quality is determined using the signal-to-noise ratio (SNR) of the links as measured by each AP.

Using the real-time statistical table view shown in Figure 5c, operators can observe the activity of the APs in the monitored network in detail. Each row in the table represents an AP in the network identified by its IP address. The statistics given by the table include: number of associated hosts, number of bytes and packets sent and received, number of packets in error and the number of dropped packets. This view is useful for quickly determining which APs are being heavily used and also to identify any APs that are not functioning properly.

While real-time visualizations provide current information about the state of a network, long term visualizations allow researchers and network operators to study trends and patterns that occur over a long period. These modules typically do not obtain information directly from the Agents. Instead they retrieve information from stored logs in a database. VISUM produces static graphs of overall network statistics continuously while monitoring a network for up-to-date views of network performance in the last day, week, month, and year. Figure 5d is a screenshot of the interactive graph visualization. Using interactive graphs the user can select a specific device to be graphed over any period of which monitored data is available. The graphs are generated dynamically from the data logged in the database. As such, the temporal range for which the logged data is processed

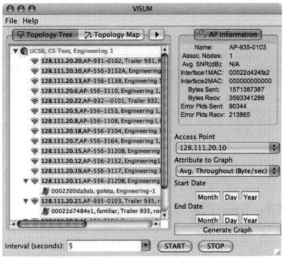

(a) VISUM real time network topology tree.

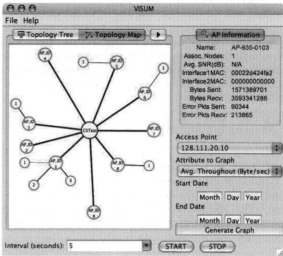

(b) VISUM real time network topology map.

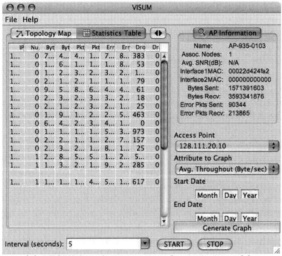

(c) VISUM real time network statistics table.

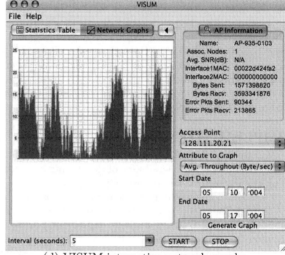

(d) VISUM interactive network graphs.

Figure 5: VISUM Data Visualization

for a given graph is also flexible. This is helpful to identify trends and patterns in network usage.

Each of the views we have implemented provides the user with a different view of the monitored information accumulated by VISUM. Although we have implemented four methods of presenting collected data, extending VISUM to provide additional methods for data visualization and presentation within the VISUM framework is straightforward. This is possible because of the object-oriented nature of Java and the modular design of VISUM.

6. CASE STUDY: UCSB COMPUTER SCIENCE DEPARTMENT WLAN

In this section we present a case study describing the use of the VISUM implementation in the UCSB Computer Science department WLAN. First, we describe the wireless network and the VISUM deployment. We then discuss the results we collected using the VISUM implementation.

6.1 Network Deployment

The UCSB Computer Science Department wireless network consists of sixteen Orinico AP-1000 APs throughout a typical office building and in several laboratories and classrooms located outside the building. This network is primarily intended as a research network and therefore access to it is restricted. Consequently, the number of users and amount of traffic on the network is low.

For this study, the VISUM agent ran continuously on one machine monitoring all sixteen APs for a period of 30 days between April 29, 2004, and May 28, 2004. We used only one agent because the size of our network is still small. Information was queried from APs at an interval of fifteen seconds. The collected data was then logged into a remote mySQL server. Initially, the configuration information, which includes description, location, and interface MAC addresses, of each AP was queried. Each subsequent log entry consisted of dynamic AP statistics, such as sent and received bytes, number of packets, error packets, dropped packets, AP uptime, and number of associated nodes, including the

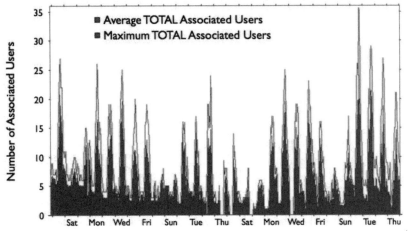

Figure 6: Average number of total associated users.

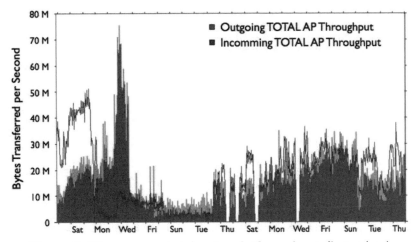

Figure 7: The average total network throughput (bytes/sec).

individual MAC addresses and traffic statistics of each associated node.

6.2 Performance Study

During the 30 day monitoring period, we recorded 237 unique MAC addresses associated at some time to the APs. Figure 6 is a graph of the average number of total associated users for the entire network over the 30 day monitoring period. Each "average total" point in the graph represents the average of 240 data samples for a period of one hour. Each "maximum total" point represents the maximum value recorded over a period of one hour. The Y-axis of the graph represents the number of associated users, and each labeled interval of the X-axis represents a duration of two days. The average total associated users for the entire network each day was only about 8 people. The graph shows that the wireless users of the network follow a weekly working schedule, in which weekdays generally see more associated users than weekends. Furthermore, daily patterns show that the number of associated users peak after 12:00 pm each day.

The average total network throughput generated by WLAN users over the 30 day monitoring period is shown in Figure 7. The Y-axis in this graph indicates the network throughput measured in bytes per second, and the X-axis is identical

to Figure 6. In this study the terms *incoming* and *outgoing* are AP-centric, i.e., incoming refers to traffic received by the wireless interface of an AP, and outgoing refers to traffic sent from the wireless interface. There are four distinct gaps in the data on the 13th, 14th, 15th and 19th of May that are attributed to power outages caused by construction of nearby buildings and maintenance reboots of the VISUM server. The peculiar characteristic of the graph, however, is the distinct lack of network activity for almost four days between Saturday May 8th and Wednesday May 12th. Re-examination of the average total users graph in Figure 6 did not indicate a significant drop in total associated users during the four day period. To further investigate the unusual behavior, we examined graphs of average throughput for individual APs. We found that the average network throughput profile of one AP in particular (AP13, shown in Figure 8) matched that of the overall network. Moreover, it was evident that AP13 was disabled during the same four day period and that the absence of traffic on this AP was responsible for the drop in network activity for the entire WLAN.

To verify our initial findings we analyzed the distribution of average throughput during the monitoring period across each of the sixteen APs. The distribution is presented in Fig-

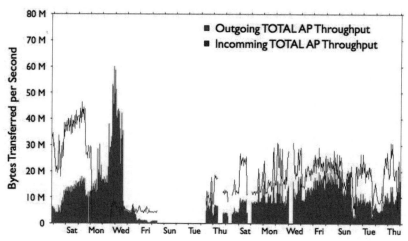

Figure 8: The average throughput for AP13 (bytes/sec).

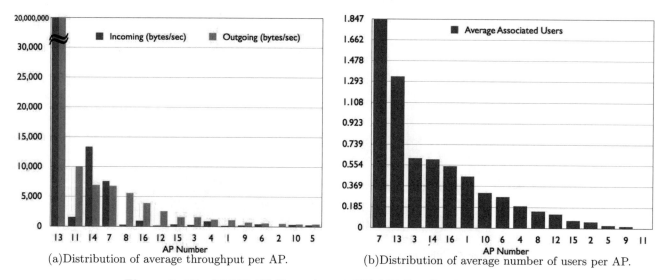

(a)Distribution of average throughput per AP.

(b)Distribution of average number of users per AP.

Figure 9: The UCSB CS Department WLAN distribution graphs.

ure 9a. The APs are ordered by average outgoing through-put from greatest to least. The graph indicates that AP13 clearly dominates all other network traffic and supports our initial findings that the traffic trends of the overall network is defined by a single AP. Figure 9b shows the distribution of the average number of associated users per AP over 30 days. Both distribution graphs suggest that the Department wireless network is over-provisioned and utilization of the network is unevenly distributed. We believe that the observations made from analyzing VISUM monitoring results are valuable for planning future expansion of the Department's wireless network infrastructure and possible integration with other university WLAN deployments.

7. CONCLUSIONS AND FUTURE WORK

Network monitoring is a critical part of many aspects of networking, in particular, protocol development, network deployment, and network management. The advancement of network technology and the maintenance of networks would be more complex and unnecessarily inefficient without effective monitoring tools to obtain network information about its state, operation, and reaction to operating conditions.

However, due to a lack of support for standardized monitoring and management protocols, heterogeneity of network components, ever-increasing network deployment, and rapid development of new network technologies, developing mechanisms to monitor networks is critical. Wireless networking is an example of a rapidly evolving network technology that needs effective monitoring for successful research, development, and deployment. Examples of practical wireless network applications that would benefit from monitoring systems are: WLAN deployment, dynamic load balancing, and network billing/accounting.

In this paper we have introduced VISUM, a framework for a distributed generic wireless network monitoring system. Using a modular architecture, VISUM collects MAC layer information from wireless network infrastructure components to provide real time views of network status. The VISUM architecture facilitates monitoring of heterogeneous wireless network components and has mechanisms to seamlessly cope with changes in network configuration. Accumulated data is logged for later analysis, or provided as real time input to a diverse set of visualization tools and high-level applications. To accommodate the monitoring of large wireless networks, VISUM's agents can be distributed among multiple hosts. Their data is easily aggregated to provide a comprehensive view of the entire network. We have demonstrated the feasibility of the architecture with an implementation developed in Java. Using the implementation we have successfully monitored the University of California, Santa Barbara Computer Science department network for a period of a month. For the presentation of monitored information, the implementation provides four distinct views of the collected information, including real time network topology and statistics tables, as well as static and interactive statistical graphs of ongoing collected data. With the capacity to monitor statistics from a varied set of wireless network components, VISUM is a flexible distributed architecture for monitoring wireless networks with diverse options for visualizing and processing the collected data.

Future work on VISUM includes a large scale deployment to monitor a more active conference environment. We also plan to develop an additional wireless network visualization that will leverage the monitoring capabilities of VISUM and incorporate geographic information of network components to provide an intuitive means to study user mobility patterns. This visualization will be useful in determining how accurately existing mobility models represent actual user behavior. Another possible extension is to provide mechanisms to facilitate management of individual wireless network components currently only monitored by VISUM. Finally, we plan to investigate techniques that can pinpoint the reason behind a certain network event, such as a congestion or an outage, through the analysis of collected information.

Ackowledgements

This work is supported by an NSF Networking Research Testbeds (NRT) grant (ANI-0335302), an NSF Infrastructure grant (EIA-0080134) and by Intel Corporation.

8. REFERENCES

[1] A. Balachandran, G. Voelker, P. Bahl, and P. Rangan. Characterizing User Behavior and Network Performance in a Public Wireless LAN. In *Proceedings of ACM Sigmetrics*, Marina Del Ray, CA, June 2002.

[2] J. Case, M. Fedor, M. Schoffstall, and J. Davin. A Simple Network Management Protocol. Internet Engineering Task Force, RFC 1067, August 1988.

[3] F. Chinchilla, M. Lindsey, and M. Papadopouli. Analysis of Wireless Information Locality and Association Patterns in a Campus. In *Proceedings of IEEE Infocom*, Hong Kong, March 2004.

[4] C. Cooper, J. Zeidler, and R. Bitmead. Modeling Dynamic Channel Allocation Algorithms in Multi-BS TDD Wireless Networks with Internet Based Traffic. In *Proceedings of IEEE Vehicular Technology Conference*, Milan, Italy, May 2004.

[5] D. Kotz and C. Newport and C. Elliott. The Mistaken Axioms of Wireless-network Research. In *Technical Report TR2003-467, Dept. of Computer Science, Darmouth College*, July 2003.

[6] P. Gummadi, S. Saroiu, and S. D. Gribble. King: Estimating Latency between Arbitrary Internet End Hosts. In *Proceedings of ACM Sigcomm Internet Measurement Workshop*, Marseille, France, November 2002.

[7] A. Jardosh, E. Belding-Royer, K. Almeroth, and S. Suri. Towards Realistic Mobility Models for Mobile Ad hoc Networks. In *Proceedings of ACM International Conference on Mobile Computing and Networking*, San Diego, CA, September 2003.

[8] K. Fall and E. Varadhan. ns notes and documentation. In *http://www-mash.cs.berkeley.edu/ns/*, 1999.

[9] D. Kotz and K. Essien. Analysis of a Campus-wide Wireless Network. In *Proceedings of ACM International Conference on Mobile Computing and Networking*, Atlanta, GA, September 2002.

[10] C. Lonvick. The BSD syslog Protocol. Internet Engineering Task Force, RFC 3164, August 2001.

[11] N. Spring and R. Mahajan and T. Anderson. Quantifying the Causes of Path Inflation. In *Proceedings of ACM Sigcomm*, Karlsruhe, Germany, August 2003.

[12] P. Rajvaidya and K. Almeroth. Analysis of Routing Characteristics in the Multicast Infrastructure. In *Proceedings of IEEE Infocom*, San Fransisco, CA, April 2003.

[13] R. Periakaruppan and E. Nemeth. GTrace: A Graphical Traceroute Tool. In *Proceedings of USENIX Large Installation System Administration*, Seattle, WA, November 1999.

[14] P. Rajvaidya, K. Almeroth, and K. Claffy. A Scalable Architecture for Monitoring and Visualizing Multicast Statistics. In *Proceedings of IFIP/IEEE International Workshop on Distributed Systems: Operations & Management*, Austin, TX, December 2000.

[15] J. Sedayao and K. Akita. LACHESIS: A Tool for Benchmarking Internet Service Providers. In *Proceedings of USENIX Large Installation System Administration*, Monterey, CA, September 1995.

[16] N. Spring, R. Mahajan, and D. Wetherall. Measuring ISP Topologies with Rocketfuel. In *Proceedings of the Conference on Applications, Technologies, Architectures, and Protocols for Computer Communications*, Pittsburgh, PA, August 2002.

[17] N. Spring, D. Wetherall, and T. Anderson. ScriptRoute: A Facility for Distributed Internet Measurement. In *Proceedings of the USENIX Symposium on Internet Technologies and Systems*, Seattle, WA, March 2003.

[18] D. Tang and M. Baker. Analysis of a Metropolitan-Area Wireless Network. In *Proceedings of ACM International Conference on Mobile Computing and Networking*, Seattle, WA, August 1999.

[19] D. Tang and M. Baker. Analysis of a Local-area Wireless Network. In *Proceedings of ACM International Conference on Mobile Computing and Networking*, Boston, MA, August 2000.

[20] V. Paxson. End-to-end Routing Behavior in the Internet. In *Proceedings of ACM Sigcomm*, Palo Alto, CA, August 1996.

[21] X. Zeng, R. Bagrodia, and M. Gerla. GloMoSim: A Library for Parallel Simulation of Large-scale Wireless Networks. In *Proceedings of Workshop on Parallel and Distributed Simulations*, Banff, Canada, May 1998.

LOCATOR - Location Estimation System For Wireless LANs

Ankur Agiwal [*]
ankur@iitd.ernet.in

Parakram Khandpur [†]
parakram@iitd.ernet.in

Huzur Saran
saran@cse.iitd.ernet.in

Department of Computer Science & Engineering
Indian Institute of Technology
New Delhi, India

ABSTRACT

With the wide spread growth of mobile computing devices and local area wireless networks, wireless network providers have started to target the users with value-added services based on the users' location information. Thus, location awareness and user tracking in indoor wireless networks has become an increasingly important issue. To address the issue, we have developed LOCATOR, a radio frequency based system for location estimation of users in indoor wireless networks. Our system works by building a radio map of the network site, which involves taking signal strength samples at various points in the wireless network, and then using this radio map, the system estimates the user's current location from the value of his current observed signal strength. Although the use of an RF-based technique for location estimation is not new, but LOCATOR is unique in the way it builds and manages the radio map to process location queries. Moreover, in addition to using ideas of probability to model the problem, we have devised a multi-level clustering based algorithm that work in conjunction with an interpolation scheme for a more efficient and accurate location estimation. We have tested our system on 802.11b wireless network testbeds and have been able to achieve an accuracy of location estimation to within 4 feet of the actual user location with 90% probability.

Categories and Subject Descriptors:
C.2.1 [COMPUTER-COMMUNICATION NETWORKS]: Network Operations – *Network monitoring, Network management*

General Terms: Algorithms, Management, Measurement, Performance, Experimentation

---

[*]Now a graduate student at the Department of Computer Science, UNC, Chapel Hill

[†]Now a graduate student at the Department of Computer Science, Stanford University

Keywords: location-aware services, user location, clustering, location management, wireless LAN, interpolation

1. INTRODUCTION

The widespread deployment of wireless local area networks and the increasing popularity of light-weight mobile computing devices has lead to an increased interest in location-aware applications and services. The goal of these applications and services is to enable the user to interact more effectively with his environment. Examples of such value added service include, but are not limited to, services like displaying the map of the immediate surroundings and guiding a user inside a building. A great deal of research has, thus, focused on developing services and architectures for location-aware systems and on the problem of location estimation and user tracking in indoor wireless environments.

The techniques used for location estimation include GPS [1], infrared[3] based systems, ultrasonic[7] based systems, wide-area cellular[2] based systems and radio frequency (RF) [9, 11, 10, 12] based systems. Among these the RF based technique for location estimation and user tracking has gained a lot of attention due to its good accuracy and ease of deployment without the need for any specialized additional hardware.

Radio frequency (RF) based location estimation systems work in primarily two phases: radio map building phase and location estimation phase. In the radio map building phase, a large number of signal strength samples from the base stations are collected at various locations in the wireless network to build a radio map database, which is essentially a mapping from the physical space to the signal space. In the location estimation phase, the system uses a user's currently observed signal strength and the radio map to estimate his location.

In this paper, we present LOCATOR: an RF-based system for locating and tracking users in an 802.11 wireless LAN framework. However, the core idea of LOCATOR is generic enough to be used for other wireless network technologies also. The key ideas that make LOCATOR accurate and scalable are: (a) Use of clustering to fragment the radio map to improve scalability and reduce the computational cost for location estimation (b) Use of probability distributions and location-based clustering along with interpolation scheme to enhance the accuracy

We have evaluated our system on multiple 802.11b testbeds but in this paper, due to space constraints, we present the results for only one of the testbeds. The experimental results of LOCATOR are very encouraging and provide estimate of user location to within 4 feet with 90% probabilty.

The remainder of this paper is organized as follows. Section II briefly reports related work on location estimation and user tracking. Section III describes the methodology of LOCATOR. In section IV, the evaluation results of LOCATOR on our testbeds are discussed. Finally, section V concludes the paper and gives directions for future work.

2. RELATED WORK

Many systems in the recent past have attempted to tackle the problem of location estimation and user tracking in wireless networks . In this section,we briefly discuss some of the major approaches.

The Active Badge System [3], an infrared-based system, was an early and significant contribution to the field of location-aware systems. In this system, a badge worn by a person emits a unique infrared signal every 10 seconds. Sensors placed at known positions within a building pick up the unique identifiers and relay the location manager software. While this system provides accurate location information, it suffers from several drawbacks: (a) it scales poorly due to limited range of infrared and (b) it incurs significant installation and maintenance costs.

Another system based on infrared technology is described in [4]. In this system, infrared transmitters are attached to the ceiling at known positions in the building. An optical sensor on a head-mounted unit senses the infrared beacons, which enables the system software to determine the user's location. This system suffers from similar drawbacks as the Active Badge System.

There have also been several location estimation systems for wide-area cellular based networks [2]. The technological alternatives for location cellular phone users involve measuring the signal attenuation, the angle of arrival(AOA) and/or the time difference of arrival (TDOA). While these systems have been found to be promising in outdoor environments, their effectiveness in indoor environments is limited by the multiple reflections suffered by the RF signal, and the inability of off-the-shelf and inexpensive hardware to provide fine-grain time synchronization.

Systems based on the Global Positioning System (GPS) [1] have also been proposed and have been found to perform extremely well in outdoor environments. Unfortunately, buildings block GPS signals and therefore, GPS does not operate in indoor environments.

A class of location estimation systems that have recently become very popular is the radio frequency (RF) based system. These systems basically work in two phases: (a) the radio map building phase and (b) location estimation phase. These phases were briefly discussed in section 1. The main advantages of these systems are their ease of deployment without the need of any specialized hardware and their reasonably good accuracy.

RADAR [9] , developed at Microsoft Research, was the first system to use the RF-based technique for location estimation and user tracking in wireless LANs. RADAR builds a single monolithic radio map for the wireless network site and carries out a k-nearest-neighbor (k typically being 3 or 4) search in the signal space. It estimates the user's location as the mean of the physical locations of the k neighbors. This approach does not give very high accuracy and moreover, since RADAR searches the entire radio map, it has the drawback of having a very high computation cost for user location estimation.

Horus [12], another RF-based location estimation system, improves upon the performance of RADAR by using their Joint Clustering technique and clustering locations in the radio map based on the strongest signal, in case there are more than 3 base stations in range. By the use of these techniques Horus achieves a significant performance gain, both in terms of accuracy as well as computation cost, over RADAR. The main drawback of the Horus system is that its location estimation algorithm can estimate a user's location only to a location that was initially used while building the radio map. Unlike LOCATOR, which incorporates interpolation techniques for location estimation, Horus system's performance degrades significantly if the granularity of the radio map is increased.

3. LOCATOR

Our location estimation system, LOCATOR, works in two phases. In the first phase, it builds a radio map of the site where the wireless network is deployed. The radio map is a mapping between the physical locations and the observed signal strength values at these locations from various base stations. In the second phase, LOCATOR uses the radio map to estimate the location of a user, given its current observed signal strength values from various base stations. These two phases are described in greater detail in the following subsections.

3.1 Radio Map Building Phase

In this phase, a database is built that maps the physical space to the signal space. The database is build only for a few selected locations, which is then used by LOCATOR for location estimation of users as described in section 3.2.

In the following subsection, the signal strength sampling strategy is discussed, followed by the explanation of the methodology adopted for fragmenting the radio map database into clusters to reduce the computation cost while answering location search queries.

3.1.1 Sampling strategy

The IEEE 802.11b standard [6] uses radio frequencies in the license-free 2.4 GHz band in which other devices like microwave ovens, cordless phones and Bluetooth devices also operate and can, thus, be a source of interference [5]. Moreover, the phenomenon of multi-path fading also causes temporal variations in the observed signal strength. Due to multi-path, the transmitted RF signal reaching the receiver through different paths has different phase and amplitude. These different components combine and produce a distorted version of the transmitted signal. Other factors causing the temporal variations include movement of people, closing and opening of doors and other environmental changes. Due to all these reasons, the wireless channel has a highly unpredictable nature.

On carrying out an experiment to study the temporal variations in the signal strength samples collected at a fixed location from a base station, variations upto 14 dBm were observed as depicted in figure 1.

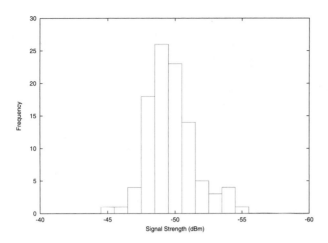

Figure 1: Temporal variation of signal strength

AP 1 (dBm)	freq.	AP 2 (dBm)	freq.	AP 3 (dBm)	freq.
S_1^1	f_1^1	S_1^2	f_1^2	S_1^3	f_1^3
S_2^1	f_2^1	S_2^2	f_2^2	S_2^3	f_2^3
S_3^1	f_3^1	S_3^2	f_3^2	S_3^3	f_3^3
S_4^1	f_4^1	S_4^2	f_4^2	S_4^3	f_4^3
$\ldots$	$\ldots$	$\ldots$	$\ldots$	$\ldots$	$\ldots$
$\ldots$	$\ldots$	$\ldots$	$\ldots$	$\ldots$	$\ldots$

Table 1: A sample frequency table for location (x, y)

Many earlier RF-based approaches like RADAR [9] have used only the mean signal strength values for location estimation. But due to the highly unpredictable nature of the wireless channel, using only the mean value of signal strength for location estimation is not sufficient for obtaining accurate results. Therefore, instead of taking only the mean signal strength value for building the radio map, a signal strength probability distribution function for various locations due to each of the base stations is defined as suggested in [12]. Assuming that the base stations operate on non-overlapping wireless channels, the signal strength probability distribution functions of the base stations at a particular location can be assumed to be independent of each other.

For simplicity only 3 base stations are assumed to be present in the network, though the model can be extended to any number of base stations. Thus, the probability of the signal strength at a location (x,y) being (q_1, q_2, q_3) due to the base stations can be expressed as:

$$P_{x,y}(q_1, q_2, q_3) = P_{x,y}^1(s = q_1).P_{x,y}^2(s = q_2).P_{x,y}^3(s = q_3)$$

where q_i is the signal strength value and $P_{x,y}^i(s)$ is the signal strength probability distribution function at location (x, y) corresponding to the ith base station.

To obtain $P_{x,y}^i(s)$, a frequency distribution table corresponding to location (x, y) is constructed, as shown in Table 1. The frequency distribution table is built by taking a large number of signal strength samples at location (x, y). For instance, $P_{x,y}^i(s = S_k^i)$ would be given by:

$$P_{x,y}^i(s = S_k^i) = \frac{f_k^i}{\sum_{j=1}^n f_j^i} \tag{1}$$

3.1.2 Clustering of radio map

The radio map database can be of a very large size depending on the size of the area where the wireless network is deployed and the granularity of the radio map database. By granularity of the radio map database, we refer to the physical closeness between two consecutive points where the signal strength samples are taken while building the radio map database. Therefore, to make the technique scalable to larger areas, the database is fragmented into location-based clusters.

Since location estimation and user tracking requires the users to periodically collect the signal strength samples from the various base stations in range and query the radio map to estimate its location, therefore, an upper bound can be assumed on the maximum distance a user can move between two such consecutive queries. This upper bound depends on the time period between two consecutive queries and the mobility pattern of the user. Therefore, to process users' queries for location estimation, instead of searching the complete radio map database, only the portion corresponding to the neighborhood of the user's previous location in the physical space needs to be searched. This strategy leads to significant decrease in the computation cost incurred for each query, leading to improved performance.

Depending on the upper bound on the maximum distance a user can move between two consecutive queries, the radio map database is fragmented into location-based clusters as shown in figure 2. Thus, if a user is in cluster (i, j) as per his previous query's estimate, then for estimating his current location the search is limited to cluster (i, j) and its neighboring clusters.

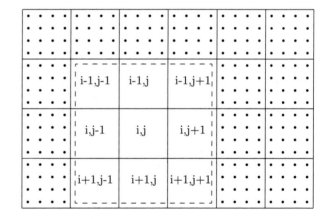

Figure 2: Fragmenting the radio map into location-based clusters

Let the upper bound on the maximum distance a user can move between two consecutive queries be d. For LOCATOR, the value of d, which is a tunable parameter, has been taken to be 15 meters. This value is based on two factors: (a) the time period between consecutive queries, which has been set to 3 seconds and (b) the maximum speed of a user, which has been taken as 5 m/s based on a survey conducted by us on the mobility behavior of users inside buildings.

Thus, with the use of location-based clustering in the radio map building phase, the computation cost greatly reduces and makes LOCATOR scalable to larger wireless networks and also allows a finer granularity of the radio map. Till now, an assumption that there are only 3 wireless base stations in the wireless network has been made. In case there are more than 3 base stations then a first level clustering based on strongest signal base station can be done as described in [12]. After the first level clustering, location-based clustering can be used without any modification.

3.2 Location Estimation Phase

In the location estimation phase, the users periodically query the radio map database with their observed signal strengths (q_1, q_2, q_3) from the 3 base stations. To estimate a user's location, LOCATOR works as follows:

1. Performs a k-best search in the 3-dimensional signal space using the previous estimate of the user's location and (q_1, q_2, q_3)

2. Clusters the k points based on their physical closeness

3. For each cluster, estimates the most probable location of the user and its likelihood

4. Chooses the location with the highest likelihood as the current estimate of the user's location

The above steps are described below in greater detail.

3.2.1 K-best search

To reduce the computation cost, the k-best search is performed only on a portion of the radio map database. If the user's previously estimated location lies in some cluster (i, j) then the search space comprises of the cluster (i, j) and its neighboring clusters only as depicted in figure 2. For k-best search, a k-nearest neighbor search in the signal space is performed using the average values of the signal strengths stored in the radio map. Two points are considered to be the closest in the signal space if their Euclidean distance is the least i.e. (q_1, q_2, q_3) and (s_1, s_2, s_3) are the closest in signal space if the $(q_1 - s_1)^2 + (q_2 - s_2)^2 + (q_3 - s_3)^2$ is minimum. Algorithms exist that perform k-best search in $O(k\,log\,n)$ time where n is the size of the sample to be searched.

3.2.2 Clustering based on physical closeness

The k candidate points $\{v_i\}_{i=1}^k$ obtained from the k-best search described in subsection 3.2.1, are now grouped into clusters based on their physical closeness.

Definition 1. The diameter of a cluster is the maximum distance between any two nodes in the cluster.

Clustering of nodes is done in a manner that each cluster has its diameter less than the maximum diameter, D. The optimum value of D is obtained empirically as discussed in section 4.4.2. Thus, if $d(u, v)$ denotes the distance between nodes u and v, then the clustering algorithm partitions the k points into m disjoint clusters $\{C_i\}_{i=1}^m$, such that

$$d(u, v) \leq D \qquad \forall\, u, v \in C_i, 1 \leq i \leq m \qquad (2)$$

Definition 2. The distance between two clusters C_i and C_j is denoted by $d(C_i, C_j)$ and is defined as:

$$d(C_i, C_j) = max\,\{d(u, v) : u \in C_i, v \in C_j\} \qquad (3)$$

Claim 1. If diameter of each of C_i and C_j is less than D and $d(C_i, C_j) < D$ then diameter of $C_i \cup C_j$ is less than D.

PROOF. Consider any edge $(u, v) \in C_i \cup C_j$
Case I: u and v belongs to same cluster. Say, $u \in C_i, v \in C_i$ then by definition of diameter $d(u, v) < D$.
Case II: u and v belongs to different clusters. Say, $u \in C_i, v \in C_j$ then by definition of distance $d(u, v) < D$. □

Algorithm for clustering:

```
1    Ci = {vi} for all 1 ≤ i ≤ k
2    S = {C1, C2, ..., Ck}
3    i = k
4    while (i > 1) do
     begin
5        find i, j  s.t. Δ = min{d(Ci, Cj)} ∀ Ci, Cj ∈ S
6        if (Δ < D)
         begin
7            Cnew ← Ci ∪ Cj
8            S = S - {Ci} - {Cj}
9            d(Cnew, C) = max{d(Ci, C), d(Cj, C)} ∀ C ∈ S
10           S = S + {Cnew}
         end
11       else stop
12       i ← i - 1
     end
```

The above clustering algorithm terminates either when all the clusters have been merged into one single cluster or when no two clusters exist such that the distance between them is less than the maximum diameter D.

Steps 1, 2 and 3 take constant time and the while loop in step 4 can loop atmost k times. Step 5 is $O(k^2)$ operation while steps 6, 8, 10, 11 and 12 are constant time operations. Steps 7 and 9 each take $O(k)$ time. Thus, the overall time complexity of the clustering algorithm is $O(k^3)$. But, k is small so number of operations are less.

3.2.3 Estimation of most probable location and its likelihood for each cluster

For each cluster, the most probable location and the likelihood of the user being at that location is calculated. For a cluster consisting of n points, Lagrange multipliers l_{1i}, l_{2i} and l_{3i} corresponding to each of the n points are calculated as described below, where (q_1, q_2, q_3) is the signal strength tuple that the user has provided in his query request.

$$l_{1i} = \prod_{1 \leq j \leq n, j \neq i} \frac{(q_1 - s_1^j)}{(s_1^i - s_1^j)} \qquad \forall\, 1 \leq i \leq n \qquad (4)$$

$$l_{2i} = \prod_{1 \leq j \leq n, j \neq i} \frac{(q_2 - s_2^j)}{(s_2^i - s_2^j)} \qquad \forall\, 1 \leq i \leq n \qquad (5)$$

$$l_{3i} = \prod_{1 \leq j \leq n, j \neq i} \frac{(q_3 - s_3^j)}{(s_3^i - s_3^j)} \qquad \forall\, 1 \leq i \leq n \qquad (6)$$

The effective multiplier at each point is calculated as average of these three multipliers.

$$\lambda_i = \frac{l_{1i} + l_{21i} + l_{3i}}{3} \qquad \forall\, 1 \leq i \leq n \qquad (7)$$

The location of the user is estimated as:

$$(X, Y) = \left(\sum_{i=1}^{n} x_i \lambda_i, \sum_{i=1}^{n} y_i \lambda_i \right) \qquad (8)$$

The likelihood, L of the user's location being (X, Y) is given by:

$$L(q_1, q_2, q_3) = \sum_{i=1}^{n} P_i(q_1, q_2, q_3) \lambda_i(s) \qquad (9)$$

where $P_i(q_1, q_2, q_3)$ is the probability that the user is located at the ith point of the given cluster. Since the base stations operate on non-overlapping channels, their signal strength probability distribution functions can be assumed to be independent of each other. Based on this assumption $P_i(q_1, q_2, q_3)$ can be expressed as:

$$P_i(q_1, q_2, q_3) = P_i^1(s = q_1).P_i^2(s = q_2).P_i^3(s = q_3) \qquad (10)$$

where $P_i^j(s = q_j)$ can be calculated from the frequency distribution table of the ith point for j^{th} base station as described in the section 3.1.

3.2.4 Choosing the location with the highest likelihood

From the previous subsection, m probable candidate locations $\{(X_i, Y_i)\}_{i=1}^{m}$ and the corresponding likelihoods $\{L_i\}_{i=1}^{m}$ for each of the m clusters are obtained. The candidate with the highest likelihood is chosen to be the estimate for the user's current location.

4. EXPERIMENTAL EVALUATION

In this section, first the experimental setup, which includes a description of the testbed and the various configurable parameters of the LOCATOR system, is described. Next, the effect of the various parameters on the performance of LOCATOR is discussed. Finally, a performance evaluation of LOCATOR against two other RF-based location estimation systems namely RADAR [9] and Horus [12] is described. For a fair performance evaluation, all three systems were tested under the same environmental conditions.

4.1 Experimental Setup

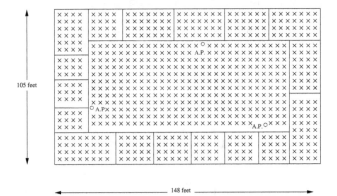

Figure 3: Experimental testbed

The testbed (figure 3), a school corridor surrounded by classrooms, is used for the performance evaluation of LOCATOR . It covers an area of approximately 15,000 square

feet and has three 802.11b wireless access points of Cisco 1200 series deployed at the positions shown in the figure. The access points operate on channels 3, 6 and 9, which are non-overlapping and therefore, do not cause mutual interference. The crosses shown in the above figure indicate the locations where signal strength samples were collected while building the radio map database.

The separation between the crosses defines the granularity of the radio map and is denoted by g. Another parameter of interest is the number of candidate points (k) to be considered for clustering in step 2 of the location estimation phase. This number defines the value for k in the k-best search that is performed in step 1. Finally, there is a parameter γ, which is the ratio of the maximum diameter of a cluster and the granularity of the radio map. Thus, the maximum diameter of the cluster that is used in step 2 of the location estimation phase is given by $\gamma.g$.

For building the radio map database, a large number of signal strength samples are recorded at each location. The number of signal strength samples recorded should be large enough so that the signal strength probability distribution function for each location can be accurately captured. For our experimental testbed, taking 150 samples at each location proved to be sufficient. Using these samples, the frequency distribution table is constructed for each of the locations. Also, the mean signal strength values at each of the locations is also recorded. The mean signal strength values are used in step 1 of location estimation phase for k-best search while the frequency distribution tables are used in step 2 for selecting the most probable location estimate for each cluster.

4.2 Effect of Parameters on Performance

LOCATOR has three control parameters: (a) Granularity of radio map denoted by g, (b) Number of candidates selected for clustering denoted by k and (c) Ratio of the maximum diameter of a cluster and the granularity of the radio map denoted by γ. In this section, the effect of variation of each of these control parameters on the performance of LOCATOR is studied. For quantifying the performance of LOCATOR, the error distance at 90 percentile is used as the performance metric i.e. for instance, if 100 observations for location estimation are made and 90 of these have error distance, which is the difference between actual position and estimated position, with in 3 feet, then error distance at 90 percentile is 3 feet.

First the effect of varying the number of candidates (k) used for clustering, on the performance of LOCATOR is studied. While varying k, g is kept fixed at 10 feet and γ at 3. As observed from figure 4, as k increases the 90 percentile error distance initially decreases and then becomes almost constant. This can be explained as follows. Initially when k is small, certain points that are physically closer to the actual user location but not that close in the signal space are not selected in the k-best search. As a result, the performance is poor. As the value of k is increased, eventually all points that are physically close to that user's current location are selected which naturally improves the accuracy. But as k is increased further, there is no improvement in the performance because all the neighboring points of the user's current location get selected after a certain threshold value of k and thus, cause saturation in accuracy.

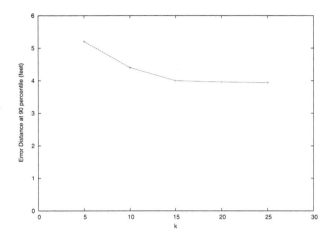

Figure 4: Effect of number of candidates, k, used for clustering

From figure 4, 15 seems to be the optimum value for k and would be used while comparing LOCATOR's performance against other radio frequency based systems in the next section.

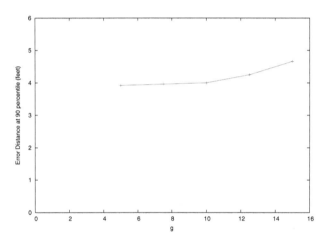

Figure 5: Effect of granularity, g

In figure 5, the effect of the granularity (g) on the performance is studied while keeping k at 15 and γ at 3. As expected, the 90 percentile error distance increases as the granularity of the radio map decreases i.e. g increases. From the figure, it is observed that initially the error distance increases at a slower rate than the rate at which g is increased. The initial slow rate of increase of the error distance can be attributed to the use of interpolation while choosing most probable locations in step 3 of location estimation phase. If interpolation were not to be used, the error distance increases linearly with the granularity. But as the granularity decreases, the cost in terms of time required to build the radio map database i.e. the time required for the radio map building phase increases. We would use the value of granularity as 10 feet as its gives the best mix of performance in terms of accuracy and the time required for radio map building phase.

Finally, the effect of variation of the ratio of the maximum diameter of a cluster and the granularity of the radio map

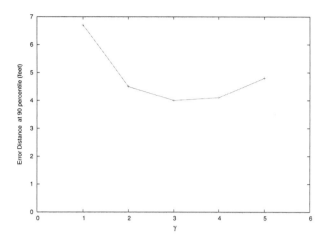

Figure 6: Effect of ratio of the maximum diameter of a cluster and the radio map granularity

(γ) on the performance of LOCATOR is studied.In figure 6, the value of γ is varied from 1 to 5 while keeping k at 15 and g at 10 feet. As observed from the figure, initially when γ is 1, the error is large which is due to the fact that there will be no clustering of points for γ equal to 1. As γ is increased, there is an increase in the performance due to clustering and interpolation of the points in the cluster. But as γ is increased further, a drop in the performance is observed. This drop in the performance can be attributed to the fact that a large value of γ can result in formation of very large clusters that comprise of points that are spaced out widely. Since all the points in a cluster are used for interpolation to estimate the most probable candidate for each cluster, as discussed in section 3.3.2.3, therefore, these distant points lead to greater errors in location estimation.

4.3 Performance evaluation against other RF-based systems

In this section, the performance of LOCATOR is compared against other radio frequency (RF) based location estimation systems, namely, RADAR[9] and Horus[12].

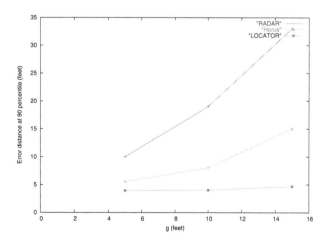

Figure 7: Performance comparison of RF-based systems on the basis of radio map granularity

First, a performance comparison of the three systems is studied on the basis of granularity of the radio map. During the evaluation, k has been kept fixed at 15 and γ at 3 for the LOCATOR system. As observed in figure 7, LOCATOR outperforms both RADAR and Horus by a significant margin and as the granularity of the radio map decreases i.e g increases, the difference in the performance becomes even more prominent. This can be attributed to the interpolation technique employed by LOCATOR to select the most probable user location for each cluster and then select the candidate with the highest likelihood as the estimate for the user location. Horus performs better than RADAR since it uses its Joint Clustering technique[12] rather than simply doing a k-nearest neighbor search as in the case of RADAR. Since neither Horus nor RADAR use interpolation, therefore the accuracy of these systems is severely limited by the granularity of the radio map. Unlike RADAR and Horus, its again due to interpolation that the performance of LOCATOR degrades gracefully as the granularity is increased.

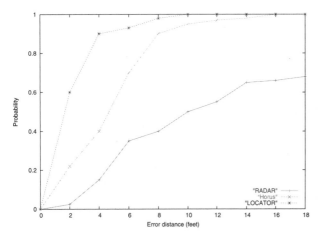

Figure 8: Error distance CDF for various RF-based systems

Next, the cumulative distribution functions (CDFs) of the error distance are presented for each of the three RF-based systems in figure 8. As observed from the figure, LOCATOR gives 90% accuracy to within 4 feet while Horus and RADAR give only 40% and 15% accuracy to within 4 feet respectively for the testbed setup that we have used.

Finally, the performance of the three systems is analyzed on the basis of the computation cost for processing location queries. For a fair analysis, all the three systems were tested on the same machine and under similar load conditions. For the purpose of comparison, the *system time* output of the Unix *time* command was used to measure the response time for location queries. Figure 9 has been plotted by taking an average of 500 location estimation queries for each of the three systems. As seen from the figure, LOCATOR significantly outperforms Horus. This can be attributed to LOCATOR's location-based clustering of the radio map due to which the entire radio map database is not searched for answering location queries, thus, leading to lower computation overhead. The performance of RADAR is just marginally worse than that of LOCATOR. This phenomenon can be attributed to two mutually nullifying factors in RADAR's search procedure. The RADAR system searches the entire radio map database for answering location queries and thus,

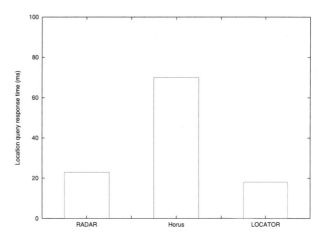

Figure 9: Computation Overhead

incurs a huge cost. This high computation cost of searching the entire radio map database is balanced out by RADAR's use of a simple, low computation cost k-nearest neighbor search for answering location queries. In conclusion, as the size of the radio map increases, the performance of LOCATOR will improve further due to its location-based clustering of the radio map database.

5. CONCLUSIONS AND FUTURE WORK

In this paper we have presented LOCATOR, a new radio frequency based location estimation system, and its performance evaluation against other RF-based systems. The effect of various control parameters on LOCATOR's performance has also been analyzed.

LOCATOR incorporates several novel ideas that lead to its superior performance. In the radio map building phase, the radio map is fragmented into location-based clusters that reduce the computation cost in the location estimation phase by a significant factor and also makes LOCATOR scalable and thus, deployable for larger wireless networks. The use of signal strength probability distribution functions, clustering and interpolation in the location estimation phase lead to a more accurate user location estimation.

The implementation results of LOCATOR on our 802.11b experimental testbed show an accuracy of 90% for error distances within 4 feet. Also, LOCATOR is a generic system that can be used for other wireless network technologies like Bluetooth etc.

In this paper, we have considered only one kind of interpolation scheme namely Lagrange interpolation. In our future work, we would like to study the effect of using other interpolation schemes and analyze the trade-off between the increase in computation cost and the increase in accuracy due to different interpolation schemes. Since the requirement, in terms of accuracy, for users' location information varies from application to application, we would also like to explore the possibility of making LOCATOR adaptive to the context in which it is being deployed. For instance, if some location-based service requires highly precise user location information then LOCATOR would switch to an expensive but more accurate interpolation scheme to provide more accurate user location.

6. REFERENCES

[1] P. Enge and P. Misra. *Special issue on GPS: The Global Positioning System.* Proceedings of the IEEE, pages 3-172, January 1999.

[2] S. Tekinay. *Special issue on Wireless Geolocation Systems and Services.* IEEE Communications Magazine, April 1998.

[3] R. Want, A. Hopper, V. Falco, and J. Gibbons. *The Active Badge Location System.* ACM Transactions on Information Systems, 10(1):91-102, January 1992.

[4] R. Azuma. *Tracking requirements for Augmented Reality.* Communications of the ACM. Vol. 36, No. 7, pp: 50-51, July 1993

[5] W. Stallings. *Wireless Communications and Networks.* Prentice Hall, first edition, 2002.

[6] The Institute of Electrical and Electronics Engineers, Inc. IEEE Standard 802.11 - Wireless LAN Medium Access Control (MAC) and Physical Layer (PHY) specifications. 1999.

[7] N. B. Priyantha, A. Chakraborty, and H. Balakrishnan. *The Cricket Location-Support system.* In 6th ACM MOBICOM, Boston, MA, August 2000.

[8] R. J. Orr and G. D. Abowd. *The Smart Floor: A Mechanism for Natural User Identification and Tracking.* In Conference on Human Factors in Computing Systems (CHI 2000), pages 1-6, The Hague, Netherlands, April 2000.

[9] P. Bahl and V. N. Padmanabhan. *RADAR: An In-Building RF-based User Location and Tracking System.* In IEEE Infocom 2000, volume 2, pages 775-784, March 2000.

[10] P. Castro, P. Chiu, T. Kremenek, and R. Muntz. *A Probabilistic Location Service for Wireless Network Environments.* Ubiquitous Computing 2001, September 2001.

[11] T. Roos, P. Myllymaki, H. Tirri, P. Misikangas, and J. Sievanen. *A Probabilistic Approach to WLAN User Location Estimation.* International Journal of Wireless Information Networks, 9(3), July 2002.

[12] M. Youssef, A. Agrawala, and A. U. Shankar. *WLAN Location Determination via Clustering and Probability Distributions.* In IEEE PerCom 2003, March 2003.

Extracting Places from Traces of Locations

Jong Hee Kang[1], William Welbourne[1], Benjamin Stewart[1], Gaetano Borriello[1,2]

[1]Dept. of Computer Science and Engineering, University of Washington, Seattle, WA 98195
[2]Intel Research Seattle, 1100 NE 45th Street, Seattle, WA 98105
{jhkang, evan, stewartb, gaetano}@cs.washington.edu

ABSTRACT

Location-aware systems are proliferating on a variety of platforms from laptops to cell phones. Locations are expressed in two principal ways: coordinates and landmarks. However, users are often more interested in "places" rather than locations. A place is a locale that is important to an individual user and carries important semantic meanings such as being a place where one works, lives, plays, meets socially with others, etc. Our devices can make more intelligent decisions on how to behave when they have this higher level information. For example, a cell phone can switch to a silent mode when the user is in a quiet place (e.g., a movie theater, a lecture hall, or a place where one meets socially with others). It would be tedious to define this in terms of coordinates. In this paper, we describe an algorithm for extracting significant places from a trace of coordinates, and evaluate the algorithm with real data collected using Place Lab [14], a coordinate-based location system that uses a database of locations for WiFi hotspots.

Categories and Subject Descriptors

I.5.3 [**Pattern Recognition**]: Clustering - *algorithms*

General Terms

Algorithms, Experimentation

Keywords

Clustering, Location-aware system, WiFi hotspots

1. INTRODUCTION

Location-aware systems are proliferating on a variety of platforms from laptops to cell phones. In these systems, locations are expressed in two principal ways: coordinates and landmarks. In coordinate-based systems such as GPS, Place Lab [14], and E911, location is specified by coordinates (latitude and longitude in this case). In landmark-based systems a location is represented as a relative proximity to one or more landmark objects. Examples of these systems include those that report the GSM cell towers within range [7] or, on a smaller scale, those that report well-known Bluetooth beacons. Location, expressed in terms of either coordinates or landmarks, is useful for many applications. For example, coordinate-based systems can be used for trip planning and navigation assistance, while landmark-based systems are useful for more local or personal applications, such as finding others that may be in the vicinity of the same landmark [6].

However, users are more interested in "places" rather than locations. A place is a locale that is important to an individual user and carries important semantic meanings such as being a place where one works, lives, plays, meets socially with others, etc. Our devices can make more intelligent decisions on how to behave when they have this higher level information. For example, a cell phone can switch to a silent mode when the user is in a quiet place (e.g., a movie theater, a lecture hall, or a place for personal reflection). A location-based reminder [3] can remind the user of what she has to carry or what she mistakenly left behind based on the user's starting point and likely destination. In a location-based to-do list application [9], the user can associate a to-do list with each place, and the application displays applicable to-do list items as the user moves about and reaches different places where they have an errand to complete. A navigation assistant application for the cognitively-impaired can guide and assist the users in reaching their destination [10].

To translate locations measured by the underlying location sensing technologies into places, we need to define the places of interest in terms of locations. For example, a user's work place can be defined as a rectangular region around her office represented in coordinates. And, if the user's current position reported by her location system is within the region (possibly with some tolerance), she is considered to be at her work place. A simple approach to define places is to define each place by hand. However, manual definition of places does not scale well. Instead, we need an approach that can automatically determine important places. These can be defined as the places where the user spends a significant amount of time and/or visits frequently. There are several parameters to consider in making this determination: duration of a visit to a place, the frequency of visits, the minimum distance between significant places, and the interaction between these three parameters.

In this paper, we describe an algorithm for extracting significant places from a trace of coordinates. In the trace, significant places are the regions where many location measurement samples are clustered together. The algorithm identifies these clusters from the trace automatically. We also evaluate the algorithm experimentally with real traces collected from Place Lab [14], a location system that uses WiFi access points' beacon messages to determine a user's location.

2. RELATED WORK

To the best of our knowledge, all previous work on place extraction with a coordinate-based location system has been done using GPS. The advantage of GPS is that it is a standardized, globally available location system that can be easily adapted for use in a variety of contexts. Potential drawbacks of GPS include its inability to function indoors, its occasional lack of accuracy due to the geometry of visible satellites, and loss of signal in urban canyons and other "shadowed" areas.

Early work on place extraction with GPS used loss of signal to infer important indoor locations. Marmasse and Schmandt [9] identify a place as a region bounded by a certain fixed radius around a point, within which GPS disappears and then reappears (as in when a user enters and leaves a building). This approach is sufficient to identify indoor places that are smaller than a certain size (e.g., a home), but does not account for larger indoor places (an office complex or convention center), and suffers from false positives (caused by the many possible outdoor GPS shadows).

A similar but more improved approach to extracting important locations is proposed by Ashbrook and Starner [1]. In that work, sets of important coordinates are identified as those at which the GPS signal reappears after an absence of 10 minutes or longer. What is more, these sets are then clustered into significant locations (i.e. places) using a variant of the k-means clustering algorithm. Through this further separation of the notions of coordinate and place, and by using a minimum time bound of 10 minutes, Ashbrook and Starner are able to overcome the place-size limitations and most of the false positives that Marmasse and Schmandt's approach suffers from. However, the use of GPS signal loss to infer place still leaves us unable to infer important outdoor places, or multiple places within a single building.

The statistical inferencing machinery used by Patterson et al. [11] and Liao et al. [8] to learn and predict daily transportation routines from GPS traces is also able to identify important outdoor places within a user's routes. Patterson et. al. use real-world knowledge of bus schedules and stop locations, along with acceleration and turning speed to infer mobile places (e.g. bus, car), as well as the location of parking lots and bus stops where users change mode of transportation. Liao et al. use mode-changes such as GPS signal loss and acceleration peaks to identify frequented locations in a totally unsupervised manner. Though identification of indoor places is still not possible, these approaches offer steps toward a more robust and complete place extraction scheme.

Recently, we have found parallel work by Hariharan and Toyama [5] that uses an approach that is very similar to ours in that they use time information to determine important places. From location histories, they first extract instances of a user spending some time in one place, which they call stays. Then, they cluster the stays and find places where one or more users have experienced a stay, which they call destinations. By using the time information in extracting stays from the location histories, they can better identify semantically important places. However, their algorithm is computationally expensive because it requires distance computations between all pairs of locations within a specified period of time to determine stays. Interestingly, they use similar tuning parameters to our algorithm and set them to almost

identical values. This gives us increased confidence in our similar algorithms.

There has also been some recent work in place extraction using a landmark-based location system. Laasonen, et al. [7] use the cells of a GSM phone network to learn important places in a user's daily routine. Their approach does not require any knowledge of network topology or even the locations of the cell towers. This approach allows place extraction over a wide area using existing infrastructure (the cellular network). However, the resolution of the derived places is very coarse (the same as that of a GSM cell – that can reach as far as a few kilometers in range although many are of much smaller range).

3. EXTRACTING PLACES

3.1 Trace of Locations

We use Place Lab [14] to collect traces of locations. Place Lab provides a way for a WiFi enabled client device to automatically determine its location. Place Lab exploits the fact that each WiFi access point periodically broadcasts its unique MAC address as part of its management beacon. Each client device holds a database that maps these addresses to longitude and latitude coordinates. When the client device receives beacon messages from nearby access points, it retrieves each access point's coordinate from the database and computes its location by averaging the locations of the access points (a simple centroid tracking scheme). The accuracy of Place Lab depends on the density and the arrangement of access points. Place Lab gives better location estimation in the areas with a high density of access points. Today, many cities and towns around the world have a high enough density of access points to provide location estimates on the order of 50-100m. Place Lab works best in urban areas – exactly the opposite of GPS which works best in open areas. More importantly, Place Lab works indoors as well where AP density is likely to be high in modern office and even residential environments.

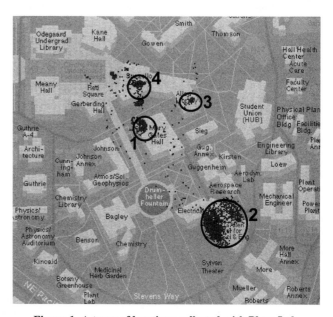

Figure 1. A trace of locations collected with Place Lab.

As with most location systems, including GPS, multiple measurements in the same location do not necessarily yield the same coordinates due to errors and variations. In Place Lab the set of access points that the client device sees in a location can vary, and consequently, the location estimate obtained by averaging the access points' locations varies as well. Thus, the important places where the user spends considerable time appear as clusters of locations in the traces. Figure 1 shows a trace of locations collected by one of the authors. The location was recorded once per second, and each location was represented as a dot in the figure. The author visited four places during the logging period, and those four places are shown as densely clustered regions in the figure.

We need to design an algorithm that will extract these significant places from the traces automatically.

3.2 Existing Clustering Algorithms

Identifying densely clustered regions from the trace is basically a clustering problem, and we first tried two popular clustering algorithms: k-means [4] and Gaussian mixture model (GMM) approach [2]. Figure 2 shows the significant places identified by these clustering algorithms.

Ideally, the system should be able to identify the evolving set of significant places by itself without input from the user. And, at the same time, the system should accurately report if the user is at one of the significant places. However, the existing clustering algorithms are not quite right for these purposes. One of the problems is that they require the number of clusters as a parameter. So, before running the clustering algorithms, the user has to specify the number of important places in advance. Although there are variations of the clustering algorithms that compute the number of clusters automatically [13], they still have other limitations. One of these is that the clusters generated by the clustering algorithms include unimportant locations. As seen in Figure 2, the clusters become unnecessarily larger when they include the intermediate and transitory locations between truly significant places. With this clustering result, the system could report that the users are in one of the significant places when they are actually merely in transit between them. Another limitation is that these clustering algorithms require a significant amount of computation and may not work well for small battery-powered mobile devices.

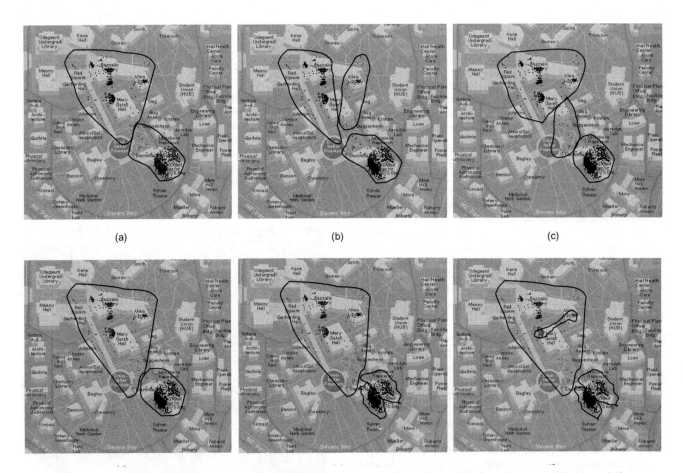

Figure 2. The clustering results from k-means and GMM. (a), (b), and (c) are the results from k-means. (d), (e), and (f) are from GMM. The number of clusters is set to 2 for (a) and (d), 3 for (b) and (e), and 4 for (c) and (f). In (c), the number of clusters is set to 4, but k-means could only find an optimal clustering with 3 clusters. Note the interesting cluster-within-a-cluster in (f).

3.3 Time-based Clustering

To overcome the drawbacks of existing clustering algorithms, an appropriate algorithm should be able to eliminate the intermediate locations between important places, and determine the number of clusters (important places) autonomously. Also, it should be simple enough to run on a simple mobile device as a background task.

The basic idea of our approach is to cluster the locations along the time axis. As a new location measurement is reported, the new location is compared with previous locations. If the new location is moving away from previous locations, the new location is considered to belong to a different cluster than the one for the previous locations. Figure 3 illustrates our approach. Suppose that the user moves from place A to place B. While the user is at place A, the location measurements are all close together (within a certain distance of each other – a parameter of our algorithm) and considered to belong to one cluster, namely, cluster a. As the user moves toward place B, the location measurements move away from cluster a. On the way to place B, a few small intermediate clusters are generated ($i1, i2,$ and $i3$). And, when the user gets to place B and stays there for a while, a new cluster (cluster b) is formed. If a cluster's time duration is longer than a threshold (the second parameter of our algorithm), the cluster is considered to be a significant place. In the figure, cluster a and cluster b are determined to be the significant places while the other clusters in between are ignored.

The algorithm is depicted in Table 1 (d and t are our distance and time threshold parameters). When a new location measurement event is generated by Place Lab, the **cluster** function is invoked. The current cluster cl is the set of location measurements that belong to the current cluster. The pending location $ploc$ is used to eliminate outliers. Even if the new location is far away from the current cluster (distance is larger than the distance threshold d), the algorithm does not start a new cluster right away with the new location. Instead, the algorithm waits for the next location to determine if the user is really moving away from the cluster or the location reading was just a spurious outlier. The *Places* contain the significant places where the user stays longer than the time threshold t.

When a new location measurement is generated from Place Lab, the algorithm compares the distance between the mean position of the current cluster and the new location with the distance threshold d. If the distance is less than d, the new location is added to the current cluster and the pending location is set to null (lines 2-3). If the distance is larger than d, the algorithm checks if there is a pending location (line 5). If there is a pending location, the algorithm closes the current cluster and checks the time duration of the current cluster (the difference between the oldest and newest locations in the cluster). If the time duration of the cluster is longer than the time threshold t, the cluster is added to the significant places (lines 6-7). Then, the algorithm starts a new cluster with the pending location and checks if the new location can be in the same cluster as the pending location (lines 8-14). If the distance between the new location and the current cluster is larger than d but there is no pending location, the algorithm set the pending location to the new location (line 16).

When a cluster is added to the set of significant places, the algorithm checks if the cluster is the same as one of the existing clusters (their centroids are within distance $d/3$ of each other). In order to identify more fine-grain places, we use a smaller threshold ($d/3$) than the one used for forming clusters (d). The smaller threshold works because the difference between the averages of the location measurements over a period of time is likely to be much smaller than the difference between individual location measurements. If the newly added cluster is close enough to one of the clusters, then the two clusters are merged.

Table 1. Time-based clustering algorithm

```
cluster(loc)
input: measured location loc
state: current cluster cl,
       pending location ploc,
       significant places Places

 1: if distance(cl, loc) < d then
 2:     add loc to cl
 3:     ploc = null
 4: else
 5:     if ploc != null then
 6:         if duration(cl) > t then
 7:             add cl to Places
 8:         clear cl
 9:         add ploc to cl
10:         if distance(cl, loc) < d then
11:             add loc to c
12:             ploc = null
13:         else
14:             ploc = loc
15:     else
16:         ploc = loc
```

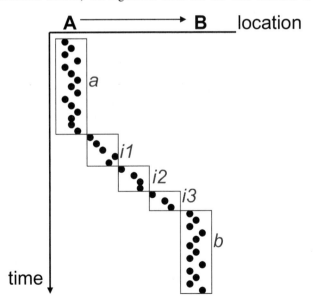

Figure 3. The illustration of the time-based clustering algorithm.

Unlike the other clustering algorithms that require all the location measurements to compute clusters, our algorithm computes the clusters incrementally as the new location measurements come in. Therefore, the significant locations can be extracted at run-time and the computation is simple enough – comparing the distance between the new location and the current cluster – to be easily supported on small battery-powered mobile devices.

Figure 4 shows the clusters generated by our algorithm. It generates four clusters from the trace. Each cluster corresponds to one of the places the user has visited. The intermediate locations between these significant places are ignored and the individual locations that make up the clusters corresponding to the significant places are the only ones shown in the figure. One interesting observation is that the shape of the bottom-right cluster is quite different from what we started with in the raw data. In the raw data, the locations in that place are clustered into two groups. But, the cluster generated by the algorithm includes only the locations in the group to the bottom-left. The user (author) was staying at the same place inside the building, but the location estimate from Place Lab was intermittent due to the variations of the signals from access points. It was moving back and forth between the two groups. It stayed in the bottom-left group for a period of time longer than the time threshold, and moved to the upper-right group for time periods shorter than the time threshold. Therefore, the upper–right group of locations were dropped by the time threshold.

The number of clusters and the size of each cluster depend on the two parameters: d and t. The distance threshold determines the size of the clusters. If a new location is further than d away from the mean of the current cluster, the new location is considered to belong to a new cluster. Thus, if d is smaller than the variations of the measurement, the algorithm may miss some significant places with high variations. In such places, even if the user stays in the same place, the location measurement can often vary a lot and the

algorithm starts a new cluster accordingly. In this case, the algorithm generates several fragmented clusters with short time duration instead of one cluster with long time duration. These fragmented clusters with short time duration are filtered out by the time threshold. If d is too large, the size of clusters may become large and absorb the intermediate locations. Also, small adjacent clusters may be collapsed into one big cluster.

The time threshold t determines the number of significant places. Only the clusters with longer time duration than t are added to the set of significant places. If t is too small, some unimportant clusters can be included in the set of significant places. On the other hand, if t is too large, we may miss some significant places that the user stops at for a time less than t.

The graphs in Figure 5 show the number of significant places found for different time and distance thresholds for two different traces of one of the authors' daily life. In both traces, there appears to be a noticeable knee in the curve between 20 and 30m. Below 20m, there are too many short duration clusters generated. Above 30m, the resulting number of clusters is quite stable. Thus, we choose the distance threshold value to be between 30m and 50m. Eventually, we will want to determine this value automatically as well and believe we will be able to do so with a few days of training data.

For the time threshold, for both traces, the graph becomes flat when the value of t is longer than about 300 seconds (for values of $d > 30$m). If the time threshold is shorter than 300 seconds, some unimportant clusters can be chosen as significant places. Therefore, we chose the time threshold to be at least 300 seconds. Depending on the user's preference, the time threshold can be set to a larger value. Ashbrook and Starner [1] did not have a clear knee in their curve and chose a value of 600 seconds somewhat arbitrarily. We believe the reason we have a more pronounced knee is the more continuous nature of our location readings in both indoor and outdoor settings.

4. EXPERIMENTAL EVALUATION

To evaluate the effectiveness of the proposed place extraction scheme, we must first decide on a criterion for evaluation. Intuitively, a place extraction scheme should be judged on how well it identifies the locales that a user deems important. The goodness of our results then, can be measured both in terms of *accuracy*, the lack of erroneously identified places, and *completeness*, the fraction of all places correctly identified. To this end, we run our time-based clustering algorithm on location traces of users' daily activity, and assess the results with both user-composed logs of the places visited (or "place logs") and map visualizations as ground truth.

4.1 Trace Collection

As noted in section 3.1, our location traces were collected using Place Lab. We used Place Lab's simple "centroid tracker", taking samples and logging location once per second. Traces were collected with wireless mobile devices during the daily activities of the first and second authors – corresponding activity logs were also composed at the end of the day. As the two authors typically stay within the Seattle city limits, and as most of this area is covered by the Place Lab AP database, there were no problems with location data being unavailable.

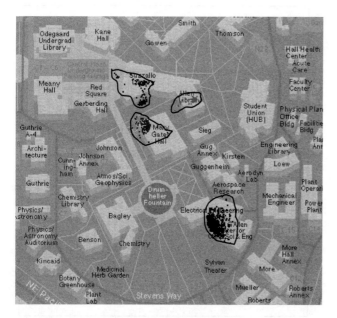

Figure 4. The significant places extracted by the time-based clustering algorithm.

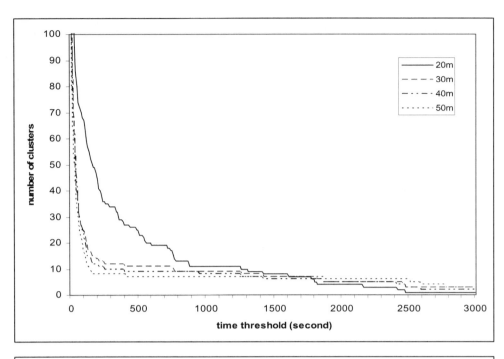

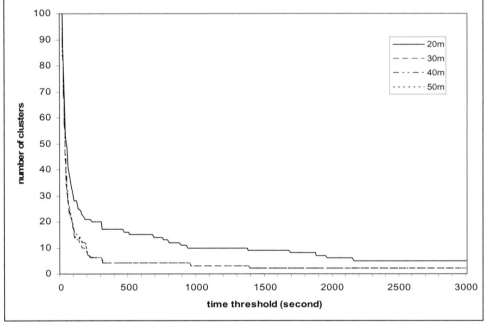

Figure 5. The number of significant places found for different distance and time thresholds.

For the purpose of evaluation, we chose one representative trace from each person. The first trace, over a small area, is of an author's daily errands around the university campus with a total duration of about 2 hours. The trace log was started when the user was in his office, place 1 in Figure 6(a). After about 10 minutes in his office, the author left to go home. On his way off campus, the author ran errands in five buildings across campus (places 2 through 6), staying 9 to 20 minutes in each place.

The second trace is of an author's daily movement between home, work, lunch, school, and a friend's house with a total duration of about 12 hours. The trace starts at the author's home (place 1 in Figure 7(a)) in the morning. After about 30 minutes, he headed to his place of work (place 2). At work, he attended a meeting in a conference room in one corner of the building, and spent the rest of the time at his desk in the other corner. After a few hours, he left to attend two meetings in a building on campus (place 3) – each meeting was held in a different room. After the second

Table 2. Detailed description of places visited in the first trace and the duration of stay in each place.

Place and Duration	Description of the Place
place 1: (10 min)	Indoor: 3rd floor office in a building; has windows; APs internal and external to the building are visible.
place 2: (17 min)	Indoor: Lobby of adjacent building; mostly concrete; internal APs of both buildings are visible.
place 3: (9 min)	Indoor: In the 3-story high atrium of a building; APs internal to building are visible.
place 4: (15 min)	Indoor: In the middle of a 1st floor corridor of a building; APs internal to building are visible.
place 5: (20 min)	Outdoor: On a bench between two buildings; open but with trees; APs from both buildings visible
place 6: (14 min)	Outdoor: On stair between two buildings; narrow "alley"; APs from near and distant buildings are visible

Table 3. Detailed description of places visited in the second race and the duration of stay in each place.

Place and Duration	Description of the Place
place 1: (35 min)	Indoor: 2nd floor apartment; has windows; APs internal and external to apartment building are visible.
place 2: (8 hour 20 min)	Indoor: 6th floor of building; office and conference room; APs internal and external to building are visible.
place 3: (45 min)	Indoor: 5th floor of campus building; offices on east and west sides; APs internal to building are visible.
place 4: (45 min)	Indoor: At table in open-air restaurant; APs external to building are visible.
place 5: (1 hour 40 min)	Indoor/Outdoor: At outdoor shopping mall; APs in various nearby buildings are visible.
place 6: (7 min)	Outdoor: On rooftop patio of apartment building; APs from both near and distant buildings are visible.

meeting ended, he returned to his place of work. At lunch time, he went out to eat at a restaurant a few blocks away (place 4). At the end of the day, he visited a shopping mall (place 5) and his friend's house (place 6) before returning home.

We evaluate our place extraction algorithm on these two traces and present the results below. For a more detailed description of each visited place, including environmental characteristics that might affect WiFi signal, please see Tables 2 and 3.

4.2 Experimental Results

Visualizations of the raw trace data for the first and second traces are shown in Figures 6(a) and 7(a) respectively. These figures also show the places listed in each author's place log as circles labeled with a number. Figures 6(b)-(d) and 7(b)-(c) show the results of time-based clustering applied to the traces for various values of d and t. The results depicted in these figures are evaluated in terms of accuracy and completeness below.

Accuracy
The raw traces shown in Figures 6(a) and 7(a) show a large number of trace points scattered along what are obviously routes between places (e.g. sidewalks, roads). We can see in Figures 6(b)-(d) and 7(b)-(c) that for each trace and each pair of clustering parameters, the scattered points between places have been excluded from the final result. Furthermore, a comparison of the results for each trace with the authors' place logs shows that each extracted place does actually correspond to a visited place.

Completeness
The results for the first and second traces show that the completeness of a set of extracted places depends largely on the choice of parameters d and t.

For the first trace, Figure 6(b) shows the places extracted when d = 30 meters and t = 300 seconds; note that only five of the six author-identified places were found in this case. The raw data in Figure 6(a) shows that the trace points around the missing place (place 6) are scattered over a relatively wide area, this is probably due to a high variation in the set of visible APs. Thus, by increasing d to 50 meters (figure 6(d)) we can compensate for scattered location estimates and recover place 6. Alternatively, if we hold d = 30 meters and increase t to 600 seconds (Figure 6(c)), we lose places 1 and 3. This is because the author spent less than 10 minutes on errands in places 1 and 3.

In the second trace, all major places (and some sub-places) with the exception of place 6 were identified with d = 50 (d = 30 yields the same result). Similar to the case of a missing place 6 in trace 1, it is likely that place 6 in trace 2 was not extracted because the surrounding trace points were scattered by occasionally visible, distant APs. This problem is likely to be lessened as Place Lab evolves to include more sophisticated tracking and AP placement schemes. In the meantime, this problem could be avoided by using a larger d value, and by using more trace data (which would presumably include more time spent in the same place, and so give the wide area enough "weight" to be considered a place).

It is also interesting to note that our algorithm was able to make the distinction between places 1 and 2 in the first trace with d set to either 30 or 50 meters. This is surprising because in the raw trace data these places look like the same cluster. Similarly, various "sub-places" could be identified depending on the value of the d parameter. For example, in the second trace the conference room and office at place 2 could be distinguished with d = 30 or 50 meters, as could the two offices in place 3 with d = 30 meters. The latter observations support the intuitive notion that a smaller d value will increase the chance that sub-places are extracted (At the same time, it will also increase the chance to miss the places with a high variation in visible APs).

5. CONCLUSIONS AND FUTURE WORK

In this paper, we presented an algorithm for extracting significant places from a trace of coordinates. The significant places where the user spends considerable amount of time appear as clusters of locations in the trace. Although this is basically a clustering problem, the popular clustering algorithms are not quite right for this particular problem for three reasons: (1) the number of clusters is an a priori parameter; (2) the generated clusters include unimportant locations; and (3) the clustering algorithms require a significant amount of computation.

Our simple algorithm clusters the locations along the time axis and extracts the clusters without a priori knowledge of the number of clusters. In addition, the clusters generated by our algorithm are more likely to tightly bound these places and to exclude extraneous coordinates. We also showed how we determined two key parameters of our clustering approach (the distance and time threshold, d and t) which we are working on learning automatically.

We also evaluated our algorithm with real trace data collected using Place Lab [14], a coordinate-based location system that uses a database of locations for WiFi hotspots. Our initial experimental results show that our algorithm extracts the most significant places successfully.

The extracted places need to be labeled in order to have semantic meanings and to be used by other applications. We are working on automatic labeling of the extracted places using additional information such as the user's calendar and other users' labeling.

Another direction for future work is to predict a user's destination from their current location and past observations of their movements. With a slight modification, our algorithm can record the arrival and leaving time to and from the extracted places. For example, each place can have the information on what time of day, or which day of the week the user visited that place. With this information, we can better predict users' destinations as they go about their day and provide proactive assistance [12].

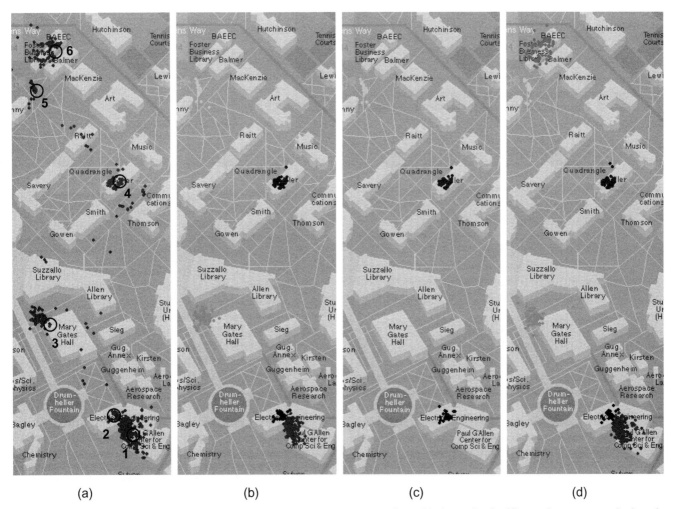

Figure 6. Visualization of the campus scale trace. (a) shows the raw trace data. (b) shows the significant places extracted when d = 30m and t = 300sec. (c) is when d = 30m and t = 600sec. (d) is when d = 50m and t = 300sec.

117

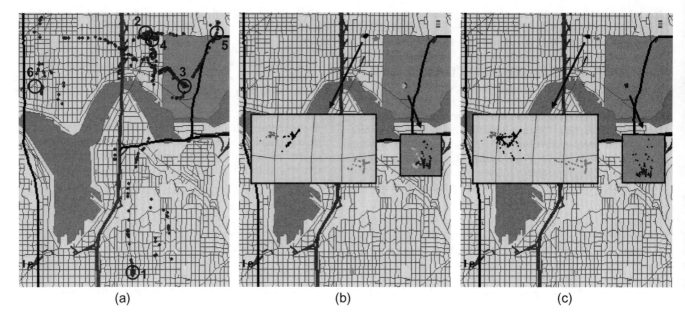

(a)   (b)   (c)

Figure 7. Visualization of the city scale trace. (a) shows the raw trace data. (b) shows the significant places extracted when $d = 30m$ and $t = 300sec$. (c) is when $d = 50m$ and $t = 300sec$. When $d = 30m$, two sub-places are extracted from place 3.

6. REFERENCES

[1] Daniel Ashbrook, Thad Starner. Using GPS to learn significant locations and predict movement across multiple users. In Personal and Ubiquitous Computing, Volume 7, Number 5, October 2003.

[2] Jeffrey D. Banfield, Adrian E. Raftery. Model-based Gaussian and Non-Gaussian Clustering. Biometrics 49, September 1993.

[3] Gaetano Borriello, Waylon Brunette, Matthew Hall, Carl Hartung, Cameron Tangney. Reminding about Tagged Objects using Passive RFIDs. To appear in Ubicomp 2004.

[4] Richard O. Duda, Peter E. Hart. Pattern Classification and Scene Analysis. John Wiley & Sons, 1973.

[5] Ramaswamy Hariharan, Kentaro Toyama. Project Lachesis: Parsing and Modeling Location Histories. To appear in the 3rd International Conference on Geographic Information Science, October 2004.

[6] John Krumm, Ken Hinckley. The NearMe Wireless Proximity Server. To appear in Ubicomp 2004.

[7] Kari Laasonen, Mika Raento, and Hannu Toivonen. Adaptive On-Device Location Recognition. In Proceedings of the 2nd International Conference on Pervasive Computing (Pervasive 2004), Vienna, Austria, April 2004.

[8] Lin Liao, Dieter Fox, and Henry Kautz. Learning and Inferring Transportation Routines. In Proceedings of AAAI-04 , 2004.

[9] Matalia Marmasse, Chris Schmandt. Location-aware information delivery with comMotion. In Proceedings of the 2nd International Symposium on Handheld and Ubiquitous Computing (HUC 200), Bristol, UK, September 2000.

[10] Donald J. Patterson, Oren Etzioni, Henry Kautz. The Activity Compass. In Proceedings of the 1st International Workshop on Ubiquitous Computing for Cognitive Aids (UbiCog '02), September 2002.

[11] Donald J. Patterson, Lin Liao, Dieter Fox, Henry Kautz. Inferring High-Level Behavior from Low-Level Sensors. In Proceedings of the 5th International Conference on Ubiquitous Computing (Ubicomp 2003), Seattle, WA, October 2003.

[12] Donald J. Patterson, Lin Liao, Kryzstof Gajos, Michael Collier, Nik Livic, Katherine Olson, Shiaokai Wang, Dieter Fox, Henry Kautz. Opportunity Knocks: a System to Provide Cognitive Assistance with Transportation Services. To appear in Ubicomp 2004.

[13] Dau Pelleg, Andrew Moore. X-means: Extending K-means with Efficient Estimation of the Number of Clusters. In Proceedings of the 17th International Conference on Machine Learning, 2000.

[14] Bill Schilit, Anthony LaMarca, Gaetano Borriello, William Griswold, David McDonald, Edward Lazowska, Anand Balachandran, Jason Hong and Vaughn Iverson. Challenge: Ubiquitous Location-Aware Computing and the Place Lab Initiative. In Proceedings of the 1st ACM International Workshop on Wireless Mobile Applications and Services on WLAN (WMASH 2003), San Diego, CA, September 2003.

Proximity Services Supporting Network Virtual Memory in Mobile Devices

Emanuele Lattanzi   Andrea Acquaviva   Alessandro Bogliolo

Information Science and Technology Institute (STI)

61029 Urbino

Urbino, Italy

{lattanzi, acquaviva, bogliolo}@sti.uniurb.it

ABSTRACT

Wireless networked embedded terminals like personal digital assistants, cell-phones or sensor nodes are typically memory constrained devices. This limitation prevents the development of applications that require a large amount of run-time memory space. In a wired cum wireless scenario, a potentially unlimited amount of virtual memory can be found on remote servers installed on the wired network. However, virtual memory access requires performance constrained and lossless data flows against terminal handover between coverage areas. In this work, we present an infrastructure aimed at providing efficient remote memory access to mobile terminals in a wireless LAN connected to a multi-hop network. The behavior of the infrastructure has been theoretically studied, implemented and tested w.r.t. various traffic conditions and handover events.

Categories and Subject Descriptors

C.2 [**Computer-Communication Networks**]: Network Operations—*Network Management*; D.4.2 [**Operating Systems**]: Storage Management—*Distributed memories, Swapping, Virtual memory*

General Terms

Management, Measurement, Performance

Keywords

Proximity service, Mobility management, Network swapping, Wireless networks

1. INTRODUCTION

Mobile wireless networks are required to support resource demanding services and applications to enable wearable and ubiquitous computing. Typical mobile terminals, such as personal communication devices and sensor nodes, have limited memory and storage resources. Storage space, if available, is mainly used for file and personal data storage, so that the amount of virtual memory available to support multi-process execution is limited unless an external memory support is provided.

To overcome this limitation, remote memory services have been proposed [10, 4] as for clusters of diskless workstations. For infrastructured networks (such as wireless sensor, cellular, home and enterprise networks), where mobile terminals (MT) are connected to a wide network through base stations (BS), an unlimited amount of remote memory space can be found on fixed computers. Remote memory service is implemented by means of a client-server mechanism similar to that supporting file sharing, where the server allocates a memory region to be used as a swapping space for a remote client.

When network memory is used for swapping the performance of the network link is critical. Moreover, the power consumption of the wireless network interface card is a primary concern for mobile terminals. Nevertheless, recent work on wireless network swapping has shown that remote memory can provide to a MT performance comparable to local storage devices at a reasonable power cost [1, 7].

The implementation of network virtual memory (NVM) services for mobile terminals, however, imposes new challenges related to mobility management. A first issue is where to locate the swap area. The choice may depend on network topology and memory availability. For performance reasons the link to the swap server must be fast and reliable. As a consequence, a viable solution would be to install the swap area as close as possible to the MT right outside the wireless network (i.e., locally to the BS). However, when the MT moves, the area content should be moved as well to provide constant QoS to the MT, while keeping the network balanced to avoid congestion on shared links.

In alternative, the swap area could be made available by a remote server connected to the Internet at one (or more) hop from the local network where the mobile terminal is connected. In this case, moving the swap content is not required, but an additional cost for slower and unreliable data transfer could be payed. Figure 1 shows the time per request taken by a remote swap server as a function of the link bandwidth.

Increasing the bandwidth of the remote link above the bandwidth provided by the wireless link doesn't improve performance. As for web services, location-aware caching

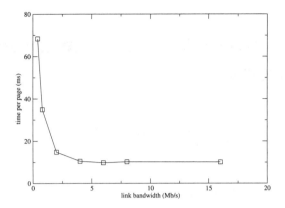

Figure 1: Time required by a remote NBD server to serve a page request as a function of the link bandwidth.

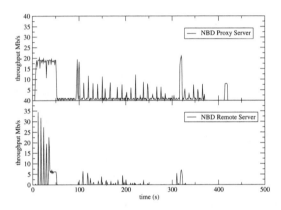

Figure 2: Swap requests to a NBD remote server filtered by a cache installed on a local server.

support should be provided to bound performance degradation in case of low bandwidth conditions. The cache can be optimized to handle most of the swapping requests locally, thus reducing network costs. The filtering effect of a local cache acting as a proxy is shown in Figure 2: only a small fraction of swap requests are forwarded to the swap server. Notice that the traffic generated by swap requests on the proxy server has peaks exceeding the bandwidth of the wireless link. This is due to two main reasons: first, the traffic on the proxy includes both the requests coming from the MT and those forwarded to the remote server, second, the BS contains internal buffers that enable traffic reshaping.

Either way, mobility management strategies should be conceived to minimize the service black-out caused by handover between different service areas.

1.1 Related Work

The problem of supporting location-aware and proximity services for mobile devices has been extensively studied. Using proxies in the wired network to support mobile devices is a well known technique especially for web browsers [6, 11]. To enhance adaptability to client movements and location awareness, the concept of proxy migration has been proposed [8] for different applications such as audio streaming and WWW browsing. However, no analysis of performance nor implementation has been given. More recently, caching

mechanisms for web data on base stations have been studied for cellular networks [6]. Based on terminal path prediction, the most popular part of the cache is transmitted to neighboring base stations. Cache relocation is performed on control area change. Control areas are groups of cells. In the case of web content this is more efficient w.r.t. to relocation on the cell handover level which is more suitable for time-critical services such as remote memory access.

For time critical services like real-time multimedia, a mechanism based on base station buffering has been proposed to minimize data loss during handover in networks based on Mobile-IP [12]. Being tailored to Mobile-IP, this method is not applicable to other networks (802.11, cellular) since it requires that data flow through the *home network* where the buffering takes place.

To support location-aware services in cellular networks, the *personal proxy* concept has been introduced [5]. It collaborates with the underlying location management system to decide when and how often it should move following the MT. By means of simulations, the optimal area size for the proxy coverage has been studied through the evaluation of a cost function having handoff delay and size of data transfer as parameters. For this reason, the personal proxy model does not take cache effects into consideration.

For 802.11b wireless networks, a context caching approach has been presented to reduce handoff delays [9]. Potential next base station knowledge is exploited to pro-actively transfer station's context, which is represented by session and QoS related information. Compared to our work, this work addresses the handoff problem at a lower level (inter access point protocol), so that the migration of application-level data is not considered.

1.2 Contribution

In this work we present location-aware proximity mechanisms providing efficient support to NVM services in an infrastructured 802.11b wireless network. We focus on optimizing context migration during handoff by means of buffering and caching strategies coupled with handoff control techniques.

We explore two main alternatives: content migration mechanisms for swap areas local to the BS, and coherency mechanisms for cached remote swap areas. We outline the behavior of each mechanism, we perform comparative performance analysis based on analytical models, we describe the implementation details and we provide experimental results obtained on a working prototype.

The NVM is implemented leveraging the *network block device* (NDB) provided by Linux. Proximity mechanisms are implemented using CORBA to enhance flexibility for remote object management (the effectiveness of CORBA to implement proxy platforms in mobile environments has already been stated [11]).

In this work we do not discuss scalability issues, but we experimentally emulated bandwidth limitation effects to evaluate the efficiency of the handoff optimization mechanisms under various traffic conditions.

2. MOBILITY MANAGEMENT OF NVM SERVICES

Although the feasibility and efficiency of remote swapping across a wireless link has been recently demonstrated both

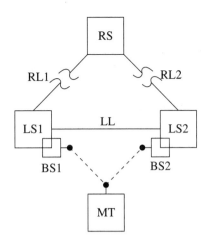

Figure 3: System architecture.

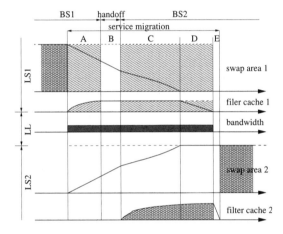

Figure 4: Timing diagram of *swap area migration* **during hand-off.**

in terms of power and performance [1], providing NVM services to a MT raises additional mobility management issues. In this section we propose several location-aware proximity mechanisms that can be used to support service mobility.

We discuss mobility management referring to Figure 3. Two base stations (BS1 and BS2) are available to provide wireless network connectivity to a mobile terminal (MT). Base stations are connected to two local servers (LS1 and LS2, respectively) connected together by means of a wideband local link (LL). A remote server (RS) is also available, connected to LS1 and LS2 by means of multi-hop remote links denoted by RL1 and RL2, respectively. We assume the MT to be initially associated to BS1 moving toward the service area covered by BS2. We want the NVM service to survive the de-association from BS1 and the re-association to BS2.

We envision two main scenarios. In the first scenario local swap areas are provided by LS1 and LS2. In this case mobility management takes care of content migration from LS1 to LS2 during handoff. In the second scenario the swap area of the MT is unique and made available by the RS. In this case LS1 and LS2 provide proximity services to enhance performance and mobility management mechanisms are required to guarantee coherency.

2.1   Local Swap Area

The swap area is local to the base station to which the mobile terminal is currently associated. This configuration provides faster and reliable access. Data transfer is not affected by traffic conditions on the wired network. The amount of traffic injected into the wired network is minimized, thus providing a good load balancing and enhancing scalability. To keep the performance level constant, the swap area content must be moved from BS to BS following terminal movements. In this context, a key issue for the infrastructure is to provide a transparent and efficient support for content migration.

The entities involved in data migration are the nbd-client on the MT, the nbd-server on the original location (LS1) and the nbd-server local to the destination (LS2). Both servers are accessible from the swap client before handoff since they are connected to the same wired network. This allows the MT to preemptively notify LS2 that it will associate to it. LS2 asks LS1 to suspend the service and starts copying the swap area. In the mean time, the MT de-associates from BS1/LS1 and re-associates to BS2/LS2. After re-association, MT may start issuing swap requests to LS2, but the requests cannot be served until the entire swap area has been transferred.

Since the swap service is suspended during swap-area migration, possibly causing service delays perceived by the MT, it is important to initiate file transfer before hand-off in order to partially hide the transfer time behind the inherent de-association and re-association time. On the other hand, depending on the size of the swap area, on the bandwidth of the local link between LS1 and LS2, and on the hand-off protocol, file transfer may take much longer than re-association, causing unacceptable QoS degradation.

2.1.1   Filter Cache

To reduce QoS degradation, a filter cache (FC) can be used that handles swap requests during content migration. The FC prevents write requests to access swap space. Write requests cause the involved pages to be loaded in cache and marked as dirty, so that future read or write requests will be performed in cache. Read requests of clean pages pass through. As a consequence, the swap area is not modified when the filter cache is enabled, allowing the migration to be performed in background. Notice that the FC is not limited in size, so that dirty pages never need to be replaced. When the entire swap area has been transferred, the MT re-associates to BS2 while the content of the FC is transferred to LS2 and used to update the content of the swap area.

In order to relax synchronization constraints between content migration and network re-association, a second filter cache can be used at LS2. This allows re-association to take place at any time during content migration.

The timing diagram of the migration mechanism is shown in Figure 4, where the sizes of swap areas and filter caches are shown as functions of time, together with the bandwidth of the local link between the two servers. Different filling patterns are used to distinguish between read-only (diagonal lines) and read-write (crossing diagonal lines) data structures. The entire process is composed of 5 phases.

Phase A. MT is still associated to BS1/LS1, FC1 is used to mask write requests while the content of SA1 is copied on SA2.

Phase B. MT de-associates from BS1/LS1 and re-associates

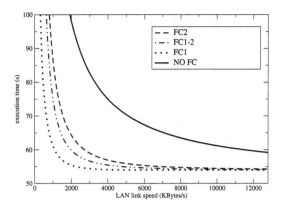

Figure 5: Swap Area Migration Policies

to BS2/LS2, the service is suspended, FC1 is frozen while the copy of SA1 on SA2 continues.

Phase C. MT is associated to BS2/LS2, FC2 is used to mask swap requests while the copy of SA1 on SA2 continues. Pages that are not found in FC2 are taken from FC1 (and, possibly, from SA1) across the local link (LL). As explained in the implementation section, this is done by means of CORBA remote method invocation.

Phase D. The content of the swap area has been completely transferred and needs to be updated. Dirty pages of FC1 are accessed through LL and written back in SA2, while FC2 is still used to mask page requests.

Phase E. Dirty pages of FC2 are written back in SA2. The service is suspended during this phase since it involves only local operations that are performed in a very short time.

2.1.2 Performance analysis

In order to analyze the performance of content migration mechanisms in different working conditions, we built a parametric analytical model providing an estimate of the execution time of a given benchmark executed across a handoff event from BS1 to BS2. The model takes in input the statistics of the swap requests generated by the benchmark without handoff (swap in and out rates), the key parameters of the handoff event (initiation instant and handoff time), the bandwidth provided by local and remote links, the size of SA and FC, a locality factor (used to tune cache efficiency to the application) and the total number of page requests generated by the benchmark. Handoff time and benchmark statistics are characterized by means of real-world experiments as described in Section 4. The analytical model is not discussed here because of space limitations.

Typical results are shown in Figure 5, where the execution time is plotted as a function of the local link speed. The solid line refers to the basic migration mechanism without FC, while all other curves refer to advanced content migrations based on FCs. In particular, the dashed line was obtained by initiating content migration and handoff simultaneously (in this case only FC2 is needed). The dot-dashed curve was obtained by initiating handoff during content migration (in this case both FC1 and FC2 are used). The dotted line was obtained by initiating the handoff after complete migration of the swap area (in this case only FC1 is used). The swap service is suspended only during handoff.

The performance speedup provided by filter caches is apparent. The best performance was obtained by performing

the handoff after SA migration. In this case, only the dirty pages contained in FC1 need to be transfered across the LL during handoff. Since the transfer time is completely hidden by the handoff, FC2 is not required thus avoiding the small black out time caused by phase E of Figure 4.

2.2 Remote Swap Area

The infrastructure described so far relies on swap areas local to each BS of the network and on high-speed links among them. Although the memory and bandwidth requirements of a single client can be easily satisfied, swap area migration gives rise to scalability issues. Using a remote swap area made accessible through the Internet may be a viable alternative. On the other hand, due to the unpredictable performance of the Internet, caching/proximity mechanisms are required to meet QoS requirements.

Caching is a very common solution for many types of web services [3]. For remote swapping, cache performance is even more critical due to the high access frequency and to the blocking nature of swap requests. We assume that each MT has a limited space (possibly much smaller than the total swap area) available on a given LS to cache swap requests. In the limiting situation where the cache contains all pages needed by a given application, the cache becomes a mirror and the RS is used only to guarantee coherency among different caches. In the typical case of caches smaller than swap space requirements, replacement algorithms need to be implemented.

Compared to the previous approach, we do not need to move the swap area from BS1 to BS2, since we consider the server far from any BS reachable by the MT, but we need to implement a cache coherency protocol. To guarantee cache coherence across handoffs cache contents must be either written back to the remote swap area, or migrated to the new location. In the rest of this section we outline the *write-back* and *migration* mechanisms.

2.2.1 Write-back mechanism

The write-back mechanism is fairly simple. When the MT de-associates from BS1, all dirty pages in the cache on LS1 are written back to the remote swap area. When the MT re-associates to BS2, a new empty cache is instantiated on LS2. The first swap request issued by the MT upon re-association causes a cache miss that cannot be served until completion of write back in order to avoid coherency problems.

The timing diagram of the write-back mechanism is shown in Figure 6.a. It consists of 3 phases:

Phase A. The write-back from LS1 is initiated while the MT de-associates from BS1/LS1 and re-associates to BS2/LS2.

Phase B. The MT is associated with BS2/LS2, but the write back is still in progress.

Phase C. Write back is terminated, the swap service is resumed, but the cache on LS2 is still partially empty.

Notice that service migration involves only phase A and B and the service is resumed as soon as the write back terminates. However, the new cache is initially empty and its efficiency is initially lower than that of the previous cache. Since the lack of efficiency is due to handoff, phase C has to be taken into account when evaluating the performance impact of handoff.

Service suspension time depends on the amount of dirty pages, which in turns depends on the cache efficiency and

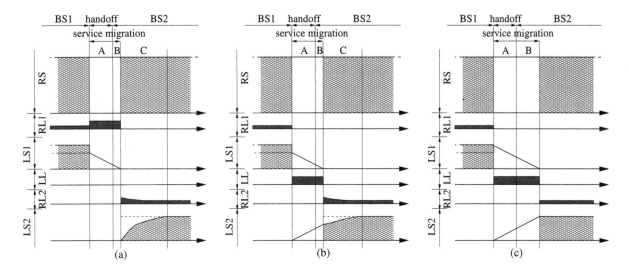

Figure 6: Timing diagrams of proximity mechanisms during hand-off: a) *write-back,* **b)** *moving-dirty,* **c)** *moving-all.*

size. Ideally, we would like the write-back time to be completely hidden by the re-association time. To this purpose, a *location-aware write back trimming* policy can be preemptively triggered by the MT when it approaches BS2 in order to keep the number of dirty pages below a given threshold (MAX). The value of MAX represents the amount of data that can be written back at full speed during the handoff time. The policy consists of triggering write back of LRU dirty pages until the number of dirty pages falls under a MIN threshold. The hysteresis is needed to avoid continuous triggering of write backs. Only the residual dirty pages need to be written back during re-association, thus making the handoff transparent to the MT.

2.2.2 Moving Cache Mechanism

Instead of writing back to the RS the dirty pages that are in cache when the handoff takes place, the same pages could be forwarded to the new LS. Cache content migration must cope with tight timing constraints and with coherency. The key advantage of cache content migration over write back is that pages are transfered across a local link (that is usually much faster than RLs) and made locally available to the MT as soon as it re-associates to the new BS.

Notice that only dirty pages need to be transferred from LS1 to LS2. After handoff, clean pages can be directly taken from the RS in case of a miss. The 3-phase process of dirty-page migration is shown in Figure 6.b.

Phase A. Dirty pages are copied from LS1 to LS2 during handoff.

Phase B. If cache migration is not completely hidden by handoff, the new cache cannot be accessed until all dirty pages have been transferred from LS1.

Phase C. The swap service is resumed, but the new cache is still less efficient than the old one since it contains only dirty pages.

The traffic induced on the remote links between each BS and the remote server and on the local link between the two BSs is also shown in the timing diagram.

The effect of transferring the entire content of the cache (including clean pages) is shown in Figure 6.c for comparison. There are two main differences from dirty-page mi-

gration. First, a larger amount of pages need to be copied from LS1 to LS2, thus possibly causing a longer service suspension. Second, as soon as the service is resumed the new cache provides the same efficiency as the old one, since it has exactly the same content.

2.2.3 Performance analysis

As for SA migration, we built a parameterized analytical model to explore the design space and perform fast comparative analysis of the proposed mobility management mechanisms.

Figures 7.a and 7.b show the execution time of a given benchmark executed across the handoff as a function of RL speed, for all mobility management policies and for two different LL speeds (namely, 24Mbps and 100Mbps). It can be observed that moving the entire cache outperforms other strategies in typical conditions, that is when the LL is faster than RL. Only in case of congestion of the LL write back policies provide better performance.

3. IMPLEMENTATION

We implemented the location aware NVM infrastructure based on the NBD provided by Linux OS. The NBD makes accessible a remote memory space by means of a server accepting read/write requests on it. The networked nature of NBD is completely transparent to other kernel components (including the `bdflush` daemon and the page fault handler that manage virtual memory) that may use the remote memory space as any other block device.

Although NVM can be transparently used by a MT thanks to the NBD abstraction provided by Linux, the implementation of the mobility management policies described in the paper gives rise to additional implementation issues that need to be carefully discussed. First, we need to be able to predict and control the 802.11b handoff mechanism in order to exploit its inherent black-out time to perform swap service migration. Second, we need to implement proximity caches and buffers. Third, we need to dynamically switch from a NBD server to another. All these issues are addressed in the following subsections.

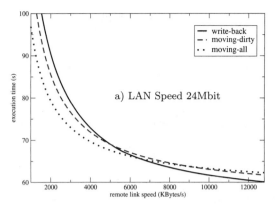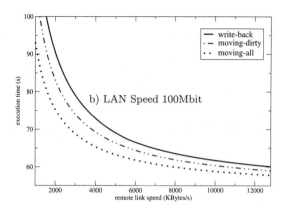

Figure 7: Performance of proximity mechanisms.

3.1 802.11b Handoff Management

The handoff process can be divided into two logical steps: *discovering* and *re-authentication*. During MT movement, the signal strength and the signal-to-noise ratio (SNR) of the current BS might fall down and cause to loose connection. Consequently the MT needs to find out if there is a new BS in range that can provide connectivity. Looking for a new BS is accomplished by the MAC layer `scan` function. There are two kinds of scanning methods defined by the 802.11b standard: *active* and *passive* scanning. While in passive scanning the station is listening for beacons to detect the presence of a BS, in active scanning mode it sends additional probe broadcast packets on each channel and waits for responses from BSs. Usually MTs use automatic active scanning to find a new BS when the signal of the current one falls under a given threshold. If a valid BS is found in range, the re-authentication phase automatically starts. Re-authentication involves the transfer of credentials and other information from the old BS (say, BS1) to the new one (BS2). If authentication is successfully completed the MT de-associates from BS1 and re-associates to BS2.

The handoff protocol described so far is completely transparent to the user. However, we need to take control of the handoff in order to exploit de-association and re-association time for swap-service migration. To this purpose we created a daemon process, running on the MT, that disables automatic active scanning and periodically scans for new BSs in range. In order to implement a user-level scanning routine we used the `mwvlan` driver for Wavelan IEEE/Orinoco that enables scanning control. For each BS in range we take SNR values in order to decide when to trigger handoff and towards which BS. We use a simple two-threshold roaming algorithm to trigger handoff as in the Sabino System [2]. When the SNR of current BS falls under the first threshold we scan for a new BS in range. When it falls under the second threshold we force re-association to the new BS. We use the time between the two thresholds to actuate preemptive swap service migration policies.

3.2 Cache Implementation

Caches local to the BS are implemented as C++ objects containing a `STL C++ map` object. The `map` object is an associative container for (key, value) couples, where value can be any C++ object. To implement a cache of swap requests, we use memory pages as values and page identifiers as keys. A memory page is an instance of a simple C++ class, called

Page, that contains page data and two flags, `r_flag` and `w_flag`, to store information about read and write activity to be used to implement replacement and migration policies.

Since caches need to be accessed from different servers during swap service migration they have been implemented as CORBA objects by specifying their remote interface using OMG IDL (*Object Management Group Interface Definition Language*). We used the *omniORB 3.0* CORBA Linux implementation and its SDK to generate the C++ stub and skeleton interface.

3.3 Server Re-association

As already mentioned, our NVM infrastructure is based on Linux NBD. The server side application provides virtual memory space by accepting read/write requests. On the client side, the NVM is made available to the Linux virtual memory management by the NBD that transparently forwards page requests to the network. The service provided by the NBD is controlled by means of a set of specific system calls. Activation consists of three steps: 1) opening a network socket to establish a connection to the swap server, 2) setting the socket into the network block device, 3) enabling the service. Once the networked device has been created and initialized, it can be chosen as swap device.

Swap service migration entails switching between two different machines providing NBD services. In principle, this can be done by changing the socket pointer into the block device. However, the Linux NBD doesn't allow the user to do so without loosing the content of the virtual memory, thus causing a kernel panic.

The key issue is making server re-association atomic in order to prevent the virtual memory management to issue swap requests during service black out. To this purpose we implemented a specific system call (NBD_CHANGE_SOCKET) locking the block device while changing the socket pointer and re-establishing the swap service. The system call can be invoked to perform swap service migration even during swap activity.

4. EXPERIMENTS AND CONCLUSIONS

In this section we briefly describe two sets of experiments conducted to characterize the inherent performance of the handoff mechanism and to test the location-aware support to NVM.

We performed our experiments using a HP's IPAQ 3600 handheld PC as MT, equipped with a Strong-ARM-1110

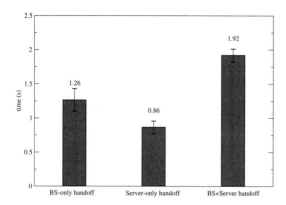

Figure 8: Experimental hand-off time.

processor, 32MB SDRAM and 16MB of FLASH. Our benchmarks were executed on the palmtop on top of the Linux operating system, Familiar release 6.0. We used a Lucent WNIC to provide network connectivity. The service infrastructure was obtained using two CISCO 350 Series base stations. Both remote and local servers were installed on 2.8 GHz Intel Pentium 4 machines, equipped with 2 Gbytes of SRAM running 2.6.1 Linux OS.

Traffic congestion on specific links was emulated by enabling the QoS option of the Linux kernel that allows the user to define band-limiting filters.

4.1 Handoff characterization

In order to characterize service migration overhead we run a set of experiments using a swap-intensive benchmark. The running task simply allocates a 2048x1024 integer matrix for a total size of 8 Mbytes. Each row of the matrix takes exactly one memory page (of 4Kbytes).

The benchmark swaps out the matrix by allocating a dummy data structure and then reads the first column, generating a page fault every time a new element is accessed. The result is a constant pattern of swap requests at the maximum sustainable rate, since no computation is performed between consecutive page faults.

We first run the benchmark with the swap area local to the BS, without performing any handoff during execution, in order to obtain a baseline performance. Then we run the benchmark again while performing the handoff operations under characterization, in order to estimate their overhead by means of differential performance measurements. Two kinds of handoff operations were characterized: *BS-only handoff*, consisting of de-association from BS1 and re-association to BS2 while using the same swap server, local to both base stations; *Server-only handoff*, consisting of disconnection from a local swap server and re-connection to a new server sharing the same swap file. The joint effect of BS and Server handoffs was also measured by performing both operations during the same run of the benchmark. Each experiment was repeated 10 times. Results are reported in Figure 8. Notice that using local swap servers accessing a shared swap file allowed us characterize inherent handoff times, independent of the mobility management policies.

4.2 Stress-test of the NVM infrastructure

We performed a stress-test of the NVM infrastructure using a simple benchmark computing the row-by-column product of two integer matrices of 2048x1024 elements. The total size of the data structures allocated by the benchmark (24Mbytes) saturates the available space in main memory. Right after memory allocation, a dummy data structure was allocated and initialized causing the matrices to be swapped out before starting computation. A regular pattern of swap requests was then generated by the benchmark to bring matrix elements in main memory.

To evaluate the performance overhead caused by the NVM infrastructure, the benchmark was first executed without handoff with three different configurations of the swap server: A) using a swap area installed on a local server, B) using a swap area installed on a remote server, C) using a remote swap area with a proxy cache local to the BS. For these experiments the bandwidth of the remote link was limited to 2Mb/s. The same benchmark was then executed with a synchronous handoff triggered right after computation of the first 200 rows. Three handoff mechanisms were tested: D) swap-area migration, E) dirty cache write back, and F) dirty cache migration.

Experimental traces showing the traffic generated by swap requests are reported in Figure 9 for each experiment. The computation time of the first 400 rows is also annotated on each graph. The first three traces, referring to a local, remote and cached NVM without handoff, show the impact of proximity on performance: accessing a remote swap server increases the execution time from 126.47 to 149.76 seconds, while caching provides a trade-off at 138.12 seconds. The filtering effect of the cache can be appreciated by comparing graphs B and C: graph B shows a baseline traffic of about 0.4Mb/s on the RL that doesn't appear on graph C.

The efficiency of the handoff mechanisms can be evaluated from the last three sets of graphs of Figure 9. Graph D refers to the swap-area migration mechanism. The graph shows the burst of traffic caused on LL by the file transfer. The dashed curve was obtained with a full-bandwidth link (100Mb/s), while the solid line was obtained by limiting the bandwidth of the LL to 20Mb/s.

Graphs E show the traffic induced on the remote links RL1 and RL2 by the write-back mechanism. Notice that, although the cache is much smaller than the total swap area (12Mbytes rather than 100Mbytes), write back takes almost the same time required to transfer the entire swap area across a full-bandwidth link. This is due to the lower bandwidth of the RL and to the lower efficiency of the write-back protocol. It is also worth noting that a large number of cache misses is generated after re-association since a new (empty) cache is instantiated on LS2.

Finally, graphs F show the traffic caused by cache content migration on RL1, LL and RL2. In this case content migration is faster, even if the remote method invocation protocol doesn't exploit the full bandwidth of the local link. The number of cache misses generated after re-association is lower than in case of write back since most of the cache content has been copied from LS1 to LS2. On the other hand, benchmark execution is slower than in case of local swap area.

As a final remark we observe that the peak bandwidth required by cache content migration on the LL is lower than 20Mb/s due to protocol inefficiency. This means that the performance of this handoff mechanism wouldn't be affected by congestion on the LL, unless the available bandwidth falls below 20Mb/s. On the contrary, swap-area migration (graph

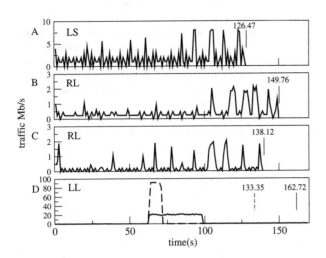
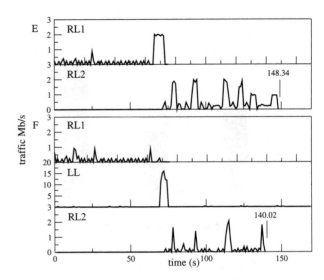

Figure 9: Link Bandwidth Occupation of Swap Area Management Policies.

D) exploits all the available bandwidth, but it's highly sensitive to traffic congestion. Hence, the comparison between the two approaches depends on the actual traffic conditions on the LL.

4.3 Conclusion

In this paper we have presented an infrastructure that provides efficient support to network virtual memory for mobile terminals. We have shown that efficiency requires proximity and we have proposed alternative strategies to keep the virtual memory pages local to the base station granting wireless network connectivity to the mobile terminal.

In particular we have focused on location-aware mobility management strategies aimed at minimizing the performance overhead caused by service migration during handoff events.

The proposed strategies have been evaluated analytically, implemented in practice and tested on a working prototype.

5. REFERENCES

[1] A. Acquaviva, E. Lattanzi, and A. Bogliolo. Power-aware network swapping for wireless palmtop pcs. In *Proceedings of the Design, Automation and Test in Europe Conference and Exhibition Volume II (DATE'04)*, page 20858. DATE, Feb 2004.

[2] F. K. Al-Bia-Ali, P. Boddupalli, and N. Davies. An inter-access point handoff mechanism for wireless network management: The sabino system. In *Proceedings of the International Conference on Wireless Network (ICWN'03)*. ICWN, Jun 2003.

[3] H. Chang, C. Tait, N. Cohen, M. Shapiro, S. Mastrianni, R. Floyd, B. Housel, and D. Lindquist. Web browsing in a wireless environment: Disconnected and asynchronous operation in artour web express. In *Proceedings of the 3rd annual ACM/IEEE international conference on Mobile computing and networking*, pages 260–269. MCN, 1997.

[4] M. D. Flouris and E. P. Markatos. The network ramdisk: Using remote memory on heterogeneous nows. *Cluster Computing, Special Issue on I/O in Shared-Storage Clusters.*, 2(4):281–293, Jun 1999.

[5] B. Gu and I. Chen. Performance analysis of location-aware mobile service proxies for reducing network cost in personal communication systems. *ACM/Kluwer Journal on Mobile Networks and Applications (MONET).*, 2004.

[6] S. Hadjiefthymiades, V. Matthaiou, and L. Merakos. Supporting the www in wireless communications through mobile agents. *ACM/Kluwer Journal on Mobile Networks and Applications (MONET).*, 7(4):305–313, Aug 2002.

[7] J. Hom and U. Kremer. Energy management of virtual memory on diskless devices. In *Proceedings of Workshop on Compilers and Operating Systems for Low Power*. COLP, Sep 2001.

[8] T. Kunz and J. P. Black. An architecture for adaptive mobile applications. In *Proceedings of Wireless 99, the 11th International Conference on Wireless Communications*, pages 27–38. ICWC, Jul 1999.

[9] A. Mishra, M. Shin, and W. Arbaugh. Context caching using neighbor graph for fast handoffs in a wireless network. *IEEE INFOCOM 2004.*, 2004.

[10] T. Newhall, S. Finney, K. Ganchev, and M. Spiegel. Nswap: A network swapping module for linux clusters. In *Proceedings of Euro-Par'03 International Conference on Parallel and Distributed Computing*, Aug 2003.

[11] R. Ruggaber, J. Seitz, and M. Knapp. π^2 - a generic proxy platform for wireless access and mobility in corba. In *Proceedings of the nineteenth annual ACM symposium on Principles of distributed computing*, pages 191–198, 2000.

[12] P. Venkataram, R. Rajavelsamy, and S. Laxmaiah. A method of data transfer control during handoffs in mobile-ip based multimedia networks. *SIGMOBILE Mob. Comput. Commun. Rev.*, 5(2):27–36, 2001.

Interoperability of Wi-Fi Hotspots and Cellular Networks*

Dilip Antony Joseph
University of California
Berkeley
dilip@berkeley.edu

B. S. Manoj
Indian Institute of Technology
Madras
bsmanoj@cs.iitm.ernet.in

C. Siva Ram Murthy[†]
Indian Institute of Technology
Madras
murthy@iitm.ac.in

ABSTRACT

The widespread deployment of Wi-Fi hotspots and wide area cellular networks opens up the exciting possibility of interoperability between these types of networks. Interoperability allows a mobile device to dynamically use the multiple network interfaces available to it so as to maximize user satisfaction and system performance. In this paper, we define three basic user profiles for the network users and demonstrate through simulation studies that dynamic switching on the basis of the user profiles of the mobile devices leads to higher network performance and increased user satisfaction. Careful design of pricing, billing and revenue sharing schemes is necessary to ensure the commercial viability of the multiple service providers involved in an inter-operable network setting. Different pricing and revenue sharing schemes are introduced and analyzed using simulation studies. We also demonstrate how load balancing can improve network performance in an inter-operable network.

Categories and Subject Descriptors

D.2.8 [**Computer-Communication Networks**]: Network Architecture and Design—*wireless communication*

General Terms

Measurement, Performance, Experimentation, Economics

Keywords

Wi-Fi hotspot, packet cellular networks, interoperability, user behaviour, pricing

1. INTRODUCTION

Recent years have seen the widespread deployment of Wi-Fi hotspots in a variety of environments such as airports, railway stations, offices, and other places of commercial interest. A number of companies offer high speed network access services through these hotspots. Cellular networks have also seen an explosive growth in the number of subscribers. For example, there are more than one billion GSM [1] subscribers and more than 50 million CDMA2000 [2] subscribers today. The cellular networks, which traditionally offered only voice calls, now offer data and multimedia services. Mobile devices of the future are likely to support both cellular and Wi-Fi interfaces.

Both cellular and wireless LAN technologies have their pros and cons. The question naturally arises – Is it possible to combine the advantages of both technologies to form a unified system? The answer is YES – Interoperability is the key. Consider the following common scenario involving a typical business executive. While traveling to his office by road every morning, the executive retrieves his email and accesses the Internet via a CDMA PCMCIA card fitted on his laptop. As he walks into his office, while a big file download is still in progress, the system transparently switches to the high speed wireless LAN (WLAN) available in the office building. The wireless LAN offers a much higher bandwidth than the cellular network and is also less expensive (often free) to access. Thus switching to the wireless LAN results in an enhanced Internet surfing experience for the executive as well as in substantial cost savings. Later in the day, if the office WLAN becomes very loaded (for instance, due to a large number of users simultaneously connecting to it), the system may again switch to the CDMA network to get a higher bandwidth, if the *user profile* of the executive demands it. All this switching happens transparent to the user – he enjoys smooth network connectivity at all times in accordance with his requirements.

1.1 Advantages of Interoperability

Interoperability allows a mobile device to dynamically use the multiple network interfaces available to it so as to maximize user satisfaction and system performance. For example, on a laptop fitted with both a CDMA wireless modem and an IEEE 802.11 WLAN card, the Internet can be accessed through either of the interfaces, depending on the user constraints and current network conditions.

Interoperability enables us to combine the advantages of both wireless LANs and wide area cellular networks in providing high speed connectivity inexpensively to a large number of users in a wide coverage area. The Wi-Fi Access Points (APs) provide high bandwidth and fast network access to the users in their coverage areas. The base stations (BSs) of the cellular network ensures network connectivity

---

[†]Author for correspondence.
*This work was supported by the Department of Science and Technology, New Delhi, India.

in a wide region, although at lower speeds. Wi-Fi APs also extend this coverage to places like subways and interiors of buildings, where the signal strength of the cellular network is very weak. Interoperability with wireless LANs also helps in relieving the heavy load on cellular networks, especially in crowded regions. Load balancing between the two different networks can increase the overall system throughput. Moreover, Wi-Fi APs can be set up at a fraction of the cost of installing BSs for the cellular network. An inter-operable system can also support a much larger number of users than any single network. This leads to higher revenues for all the network service providers involved.

1.2 Issues in Interoperability

Achieving interoperability between wireless LANs and cellular networks is a very challenging task. Here we briefly discuss some of the major issues involved in achieving interoperability.

- **When to Switch?**

  The manner in which a mobile device dynamically switches between the multiple network interfaces available to it greatly influences the performance and resource consumption of the system. *User Profile* based switching as a possible solution is explored in Section 3.

- **Smooth Handoffs**

  Maintaining existing network connections while switching between different network interfaces is a difficult task. Smooth handoffs involve issues of diverse addressing schemes, different packet formats and sizes, and packet sequencing across multiple networks.

- **Billing and Revenue Sharing**

  The presence of multiple service providers makes billing and revenue sharing very challenging tasks. It is essential that the schemes are so designed that the commercial viability of all service providers is guaranteed.

- **Security and Authentication**

  A mobile device can connect to multiple networks at the same time. Authentication of the users and security of data transmitted across diverse networks are not easy to achieve.

- **Load Balancing**

  Load balancing between different network architectures requires new metrics to ascertain the load and novel schemes to shift a section of users to a different network, whenever needed.

- **Implementation**

  Implementation of interoperability requires changes in both the network protocols as well as in the protocol stacks of the mobile devices. Maintaining compatibility with the existing systems and protocols while incorporating interoperability is very important.

- **Quality of Service**

  Mobile devices of the future will run applications that have stringent Quality of Service (QoS) requirements. Ensuring QoS in a system supporting handoffs between multiple networks with diverse characteristics is a very challenging problem.

- **Inter-Service Provider Agreements**

  Interoperability calls for co-operation between the different service providers on a large number of issues - for example, in billing and revenue sharing. Conflicting interests often make inter-service provider agreements hard to achieve.

The rest of this paper is organized as follows. In Section 2, we describe the interoperability system architecture and packet routing mechanisms. Section 3 describes the impact of user behaviour on interoperability. Section 4 discusses various pricing schemes and revenue sharing schemes, and also introduces the concept of load balancing by dynamic pricing. Issues in the implementation of interoperability are discussed in Section 5, while Section 6 concludes the paper.

2. SYSTEM ARCHITECTURE

The system under consideration consists of a set of Base Stations (BSs) belonging to a Packet Cellular Network that can provide wide area coverage to the Mobile Stations (MSs), and a set of Wi-Fi Access Points (APs) that provide high speed connectivity to the MSs. Each MS is assumed to support only one wireless interface, that can switch between the packet cellular and Wi-Fi modes of operation. The Wi-Fi APs are assumed to be interconnected with one another, and with the BSs by means of either a wired backbone network, or by high bandwidth point-to-point wireless links. The BSs are connected by means of a high speed wired network.

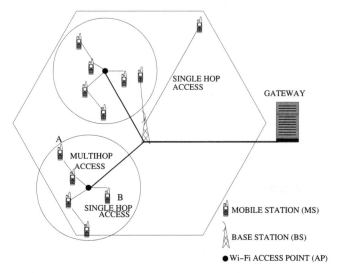

Figure 1: The Interoperability Framework

Figure 1 shows a schematic representation of the system under consideration. The BSs are placed such that the entire terrain is covered, while the Wi-Fi APs are assumed to be randomly distributed throughout the metropolitan area. The packet cellular network in our system is represented by a Single hop Cellular Network (SCN), in which the MSs are in communication with the BS on the control as well as data channels, with a transmission range equal to that of the cell radius. The Wi-Fi hotspots are considered as multi-hop relaying environments similar to the Multi-hop Cellular Networks described in [3] and [4]. The Wi-Fi AP acts as the coordinator for enabling routing and reserving bandwidth for MSs in the hotspot. The MCN architecture as described in [5] assumes a control interface of transmission range equal to that of the cell radius, and a data channel with a transmission range equal to half the cell-radius. Wi-Fi APs with multi-hop relaying is an attractive option in WLANs as it can extend the coverage of a high bandwidth AP to a much larger area. As is evident from the system architecture, each MS has the option of operating either under the control of the BS or under the control of the Wi-Fi AP.

The protocol proposed in [5] works as an infrastructure-aided source routing mechanism [6], that uses the topology information available at the BSs. We now present a brief description of the protocol in [5], and also describe some of the modifications needed to support the new inter-operable architecture. Each BS periodically generates *Beacon* messages that can be received by all mobile stations within its coverage area. An MS chooses to register with a particular BS depending on the received signal strength. It then sends the *RegReq* packet, to which the BS replies with a *RegAck* packet.

Once the registration is complete (*i.e.,* after the MS receives the *RegAck* packet), the MS will periodically generate *Beacon* messages. It updates the BS with information about the set of neighbors that are within its transmission range through *NeighUpdt* packets. Whenever there is a packet to be sent, the source MS originates a *RouteReq* packet to its BS. The BS responds with a *RouteReply* packet containing the shortest path which the MS uses to source route the packet to its destination.

In order to deal with MSs of different user profiles, both the BSs and the Wi-Fi APs periodically generate *Beacons* with transmission ranges R, the cell radius of the cellular network and r, the transmission range of the control channel of the Wi-Fi hotspot (modeled as an MCN) respectively. Each such beacon advertises the per byte transmission cost levied by the AP or BS as well as an estimate of the free bandwidth that will be available to a new MS registering with it. An estimate of the free bandwidth is obtained by dividing the total bandwidth available at the BS or AP by the total number of MSs currently registered with it. This approach assumes that the APs and BSs have a fair packet scheduling scheme running. The crucial difference between the MSs with different user requirements occurs in the registration mechanism.

The network usage that we consider in this paper is essentially one of gateway access, with one of the BSs acting as a gateway to the Internet or as a content server. This means that each MS needs to only find a route to its nearest infrastructure node (either a BS or an AP), which can then connect to the gateway by means of the backbone network. For MSs registered to the Wi-Fi AP, the routing mechanism proceeds in a similar fashion to the MCN routing protocol discussed above. However in the case of multi-hop wireless LANs, the problem of network partitions can arise, especially if the node density around the AP is low. This essentially means that the MS cannot find a multi-hop path over the data channel (of transmission range $r/2$) to its AP. In such a case the AP generates a *PartitionMsg*, to indicate to the MS that it is in a partition, and thus cannot utilize the network. On receiving the *PartitionMsg*, the MS deregisters from the AP and tries to use the nearest BS, so that connectivity is not lost.

3. USER BEHAVIOUR

A mobile station (MS) often finds itself in the coverage of multiple networks at the same time – for example, that of a cellular network and that of a Wi-Fi hotspot. The MS can choose to connect to any of the available networks. The behaviour of the MS is driven by its resource requirements and user interests. For example, an MS engaged in a multimedia transmission will have different requirements from one which is just downloading email from a server. We associate each MS with a user profile [7] that reflects its requirements. This user profile in turn determines how the MS chooses the cellular BS or Wi-Fi AP it connects to. The following are three basic user profiles a mobile station may possess.

1. **CLASS 1 – Bandwidth Conscious User Profile**: The MSs with a Bandwidth Conscious user profile will choose to connect to the BS or Wi-Fi AP which offers the maximum bandwidth. An estimate of the free bandwidth available is sent along with each beacon packet periodically originated by the APs and the BSs. A CLASS 1 MS on receiving such a beacon determines to switch to the new BS or Wi-Fi AP if the bandwidth advertised is greater than the free bandwidth estimate at its currently registered BS or Wi-Fi AP by a threshold value. MSs with high bandwidth requirements (like those engaged in a multimedia download) possess this type of user profile.

2. **CLASS 2 – Cost Conscious User Profile**: An MS with a Cost Conscious user profile always chooses to connect to the network with the lowest transmission cost per byte. Each BS and Wi-Fi AP advertises its associated transmission cost in the beacons sent by it. A CLASS 2 MS will switch to a new network only if the cost of the new network is less than that of its currently registered network by a threshold value. An MS engaged in non-real time file downloads can possess this user profile.

3. **CLASS 3 – Glitch Conscious User Profile**: A Glitch Conscious MS has glitch free connectivity as its priority. We define a glitch as an interruption in the transmission or connectivity which occurs when an MS moves from one network to another. Thus an MS with this user profile tries to minimize the number of hand-offs it undergoes between different networks to achieve the smoothest possible transmission. This is done by remaining connected with the cellular network, which has a larger coverage area, at all possible times. An MS engaged in a voice call may use this profile.

In all the three different user profile classes, we assume that maintaining connectivity is of utmost importance to the MS. In order to maintain connectivity, an MS may connect to a network whose parameters go against its user profile. For example, a Cost Conscious MS may connect to a higher cost network when it falls outside the coverage area of the low cost network to which it is currently connected.

The user profile of an MS affects MS's behaviour, the resource consumption of the network and the traffic patterns, as it moves across the terrain. This behaviour is illustrated in Figure 2, in which an MS moves from point A to point E along the dotted line shown. The total bandwidth available at the base station BS1 is 1 Mbps while that at the two access points AP1 and AP2, are 11 Mbps each. In the scenario depicted here, we assume that the free bandwidth available at AP1 is much less than that at either BS1 or AP2 due to a large number of MSs registered to AP1. The free bandwidth available at AP2 is greater than that at BS1. Also, the per byte transmission cost associated with BS1 is considered to be four times that of either AP1 or AP2. We now describe the behaviour of the MS for each of the three different user profile classes it may posses.

- **CLASS 1 MS:** The bandwidth conscious CLASS 1 MS always tries to be registered to the network offering the maximum free bandwidth. It can be seen from Figure 2 that the MS registers with the sole AP accessible to it at the beginning of its journey (Point A). At point B, the MS comes under the transmission range of BS1 also. Since BS1 has more free bandwidth than AP1, the MS will switch over to BS1 and will remain registered with it till the point D. On entering the range of AP2 at D, the MS switches over to AP2, although it is still in the range of the BS. This is because AP2 offers a higher amount of free bandwidth than BS1. The MS remains with AP2 till the end of its journey.

- **CLASS 2 MS:** The cost conscious CLASS 2 MS tries to register with the least cost network at all times. After starting its journey from point A, the MS remains registered to AP1 till the point C. It must be noted here that the MS does not switch over to BS1 after it enters BS1's transmission range at point B. This is because the transmission cost associated with BS1 is higher than that associated with AP1. At point C, the MS goes out of the range of AP1 and is thus forced to register with the higher cost BS1 in order to maintain connectivity. On reaching the point D, the MS registers with the lower cost AP2 and remains with it till the end of its journey.

- **CLASS 3 MS:** The glitch conscious CLASS 3 MS tries to minimize the number of glitches in its connection by registering with the larger range cellular network at all possible times. The MS remains registered with AP1 between points A and B. However, once it enters the range of BS1 at point B, it switches over to the BS and remains with it till the end of the journey at point E.

3.1 Impact of User Behaviour

We have studied the impact of user profiles on the network throughput, per byte access costs and number of glitches suffered by the mobile nodes through extensive simulations in GloMoSim[8]. Table 1 gives the default parameters used across all simulation experiments described in this paper, unless specifically indicated otherwise. Random waypoint mobility model was used across all simulations.

Figure 3 shows the variation in Packet Delivery Ratio (PDR) of the three classes of mobile nodes as the mobility is varied from 2 m/s to 20 m/s. We observe the surprising result that, although the bandwidth conscious nodes attain the highest PDR at low mobility, at high values of mobility the glitch conscious nodes obtain a PDR higher than that of even the bandwidth conscious users. This can be attributed to the fact the glitch conscious users remain registered to

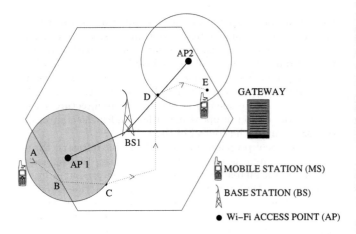

Figure 2: Behaviour of the mobile stations with different user profiles as they move across the terrain

the larger coverage cellular BSs as against the bandwidth conscious users, which suffer a very large number of glitches while remaining registered to the smaller coverage but high bandwidth APs at high mobility.

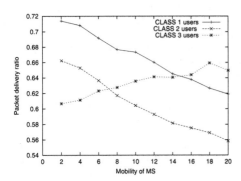

Figure 3: Packet Delivery Ratio versus Mobility of the MSs

Figure 4 shows that the glitch conscious users incur the highest per byte transmission costs. We can also observe from Figures 3 and 4 that the bandwidth conscious users incur costs comparable to or even lower than that of the cost conscious users. This leads to the important result that it is possible for the bandwidth conscious nodes to attain high PDRs at very low costs. This result follows from the fact that the Wi-Fi APs offer high bandwidth at a low price.

As shown in Figure 5, increasing mobility is also accompanied by an increasing number of glitches for all three classes of MSs. We observe that the glitch conscious nodes suffer the least number of glitches. Bandwidth conscious nodes switch every time a beacon advertising a free bandwidth higher than that of its currently registered BS or AP is received. This results in these type of nodes suffering a much larger number of glitches than the cost conscious nodes, which switch between APs and BSs only on the basis of cost. It is possible to reduce the number of glitches suffered by bandwidth conscious users by fixing a *bandwidth switch threshold*, as shown by Figure 6. However, simulations also

Table 1: Default Simulation Parameters

Parameter	Value	Parameter	Value
Terrain X range	4020m	Beacon period	1s
Terrain Y range	5220m	Bandwidth BS (Control)	1 Mbps
Number of cells	11	Bandwidth BS (Data)	1 Mbps
Cell radius	1km	Bandwidth AP (Control)	1 Mbps
Transmission range BS	1km	Bandwidth AP (Data)	11 Mbps
Transmission range AP (Control)	250m	Transmission range AP (Data)	125m
Transmission cost per byte (BS)	4 units	Number of MSs	600
Transmission cost per byte (AP)	1 unit	Mobility of the MSs	10 m/s
Mean inter-packet arrival time	2000	Number of APs	40

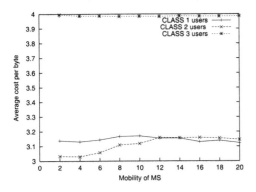

Figure 4: Average Per Byte Access Cost vs Mobility of the MSs

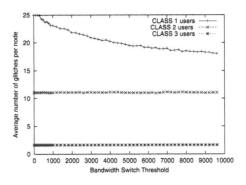

Figure 6: Average number of glitches suffered by an MS vs Bandwidth Switch Threshold

indicate that the PDR or CLASS 1 MSs does not improve in spite of the lower number of glitches suffered. The gain in PDR due to lower number of glitches is offset by the disadvantage of missing out high free bandwidth APs or BSs due to large bandwidth switch thresholds.

present in the system is low, the highest PDR is achieved by the glitch conscious users due to the minimal load at the cellular BSs. As the number of nodes is increased, the PDRs of all three types of nodes decrease, but the highest PDR is attained by the bandwidth conscious nodes. This result is captured by Figure 8.

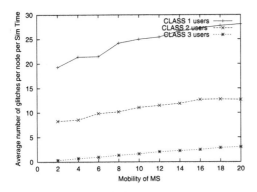

Figure 5: Average number of glitches suffered by an MS vs Mobility of the MSs

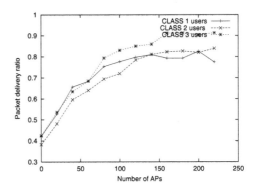

Figure 7: Packet Delivery Ratio vs Number of APs

Figure 7 show that increasing the number of Wi-Fi APs in the system leads to higher PDR until a limit, after which the performance of the system degrades due to increased congestion. This is because all the Wi-Fi APs operate in the same frequency channel. Frequency planning among the APs located close together may reduce the degradation in system performance. When the number of mobile nodes

The Wi-Fi APs are usually single hop environments. When the APs are modeled as multi-hop relaying environments, it is found (Figure 9) that glitch conscious nodes attain the highest PDR. This is due to the large number of network partitions (lack of a multi-hop path between the node and AP) suffered by the bandwidth conscious and cost conscious users when they are connected to the multi-hop Wi-Fi APs.

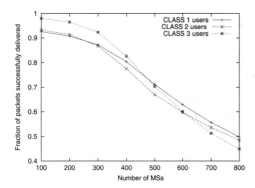

Figure 8: Packet Delivery Ratio vs Number of MSs

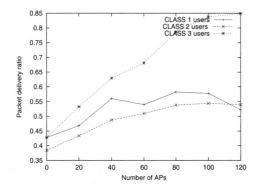

Figure 9: Packet Delivery Ratio vs Number of APs (APs in multi hop mode)

4. REVENUE SHARING

An inter-operable network setting involves interaction between multiple service providers. It is essential that the commercial viability of all the services providers involved in the system is guaranteed. Thus pricing and revenue sharing are very important aspects of an inter-operable network.

4.1 Pricing Schemes

Flat-rate pricing and *volume-based pricing* are two major pricing schemes that can be employed. Flat rate scheme is the currently existing and popular scheme. In flat-rate pricing, a user is permitted to utilize the network services for a specified period of time at a fixed price without any restrictions on the bandwidth consumed. The volume-based pricing approach charges a user based on the amount of data which he/she transacted over the network. In addition to both these schemes, business establishments can provide Wi-Fi network services for free, as a value addition to customers visiting their premises for core business activities.

A flat rate scheme is attractive for the network users (MSs) at low network loads. We can see from Figure 10 that average per byte cost incurred by the MSs increases as the number of MSs in the system is increased. In the presence of a large number of MSs, the number of packets that are successfully sent or received by the MSs decreases as a result of the increased congestion in the network. This causes the increase in the average per byte cost incurred by the MSs.

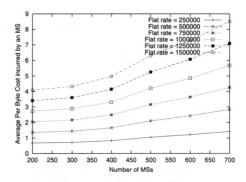

Figure 10: Average per byte cost incurred by an MS versus the number of MSs in the system at various flat rate prices.

A higher flat rate implies higher revenues for the SP as well as a higher average per byte cost for the MSs. We can see from Figure 11 that volume based pricing results in a higher revenue than flat rate schemes when the number of MSs in the system is low. But in this case, the average per byte cost incurred by the MSs in the volume based scheme is higher than that in the corresponding flat rate scheme. However at particular values of load (for example with 450 MSs in the system), flat rate schemes (for example, flat rate $= 10^7$) were found to generate both higher revenues for the service providers, as well as charges a lower average per byte cost to the MSs, than volume based schemes.

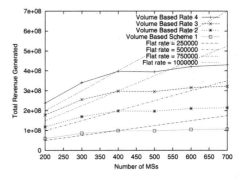

Figure 11: Comparison of total network revenue under various flat rate and volume based schemes (per byte rate = 4, 3, 2, and 1)

4.2 Revenue Sharing

Revenue sharing models describe the way in which money paid by the network users is split among the various service providers (SPs). The revenue sharing scheme used in a particular system directly influences the revenue obtained by the different service providers, and in turn determines their profitability and commercial viability. In a *Fixed-fraction sharing model*, the total revenue to be shared among the service providers is fixed apriori. In a *Volume-based sharing model*, the revenue obtained by a particular service provider depends on the volume of data transacted by that service provider.

In a simple volume based revenue sharing scheme, the WAN SP and Billing Agency treat the Wi-Fi AP as an intermediate node that participates in the forwarding of data. The Wi-Fi AP is reimbursed the amount β by the WAN SP for every byte of data it has forwarded. This scheme is illustrated in Figure 12. The value of β has to be decided in such a way that the WAN SP's revenue does not fall below a minimum threshold, and at the same time, the Wi-Fi SPs also get a fair share of the generated income. Since this model pays the APs on the basis of the traffic that they have transmitted instead of equal sharing among the APs, it can also be used when each AP is operated by a different Wi-Fi SP.

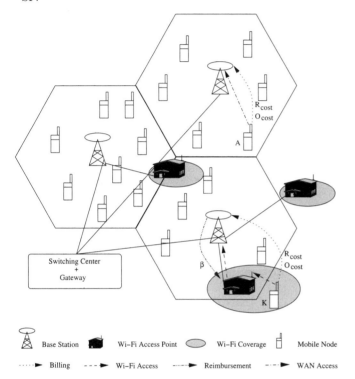

Figure 13 shows illustrates the above described β schemes. Beta schemes 1 and 2 represent the Constant β schemes with the value of the constant equal to 2.5 units and 1.5 units respectively. Scheme 3 is Level-based, while Scheme 4 represents the Continuous Function scheme $f(t) = 3.9e^{-10^{-5}t}$, where t is the total number of bytes transacted through the Wi-Fi AP.

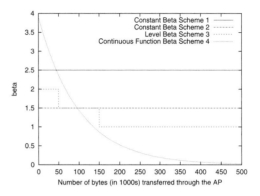

Figure 13: Variation of β with total number of bytes transferred in the various β schemes

Figure 14 shows that, among all the four schemes, Scheme 1 generates the maximum revenue for the APs. However, this scheme also results in the lowest revenue for the BSs, as seen in Figure 15. We can also observe that schemes 2, 3, and 4 generate approximately the same amount of revenue for the BSs. However in Scheme 2, at low number of MSs in the system, the revenue reimbursed to the APs is very low. This revenue may not be sufficient to keep the operation of the AP commercially viable. The advantage of the level based and continuous function schemes is that the APs are reimbursed sufficient revenue to maintain commercial viability even at low loads without decreasing the BSs' revenue by a large amount, as is evident from Figure 14.

Figure 12: Illustration of a simple revenue-sharing scheme

We compare the following three schemes for fixing the value of β:

- **Constant Beta**

  This is the trivial scheme in which the Wi-Fi AP receives the constant amount β for each byte of data forwarded by it.

- **Level-based Beta**

  In this scheme, the value of β varies in a step-like fashion with the total number of bytes forwarded. For example, $\beta = b_1$ for the first x bytes of data forwarded, $\beta = b_2$ for the next y bytes and $\beta = b_3$ for the remaining.

- **Continuous Function Beta** This scheme is a variation of the Level-based Beta Scheme in which the value of β varies continuously for each byte of forwarded data, i.e., $\beta = f(t)$ for the tth byte of data forwarded. $f(t)$ is commonly taken to be a negative exponential function.

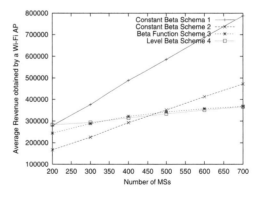

Figure 14: Effect of the different β schemes on the average revenue earned by an AP

4.3 Balancing Load by Dynamic Pricing

Differentiated network access costs provide an opportunity to balance the load across the various APs and BSs. Dynamically increasing the cost associated with a BS or an

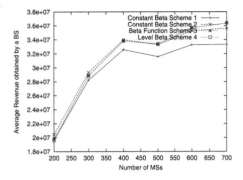

Figure 15: Effect of the different β schemes on the average revenue earned by a BS

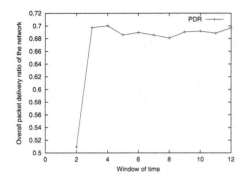

Figure 16: Variation in Packet Delivery Ratio with simulation time

AP will cause MSs with the cost conscious user profile defined in Section 3 to switch away to lightly loaded (and hence lower cost) APs or BSs in their vicinity. This balancing of load results in a higher network throughput. This scheme also has the added advantage of decreasing the average per byte access cost incurred by the cost conscious MSs.

Global Average Balancing and *Local Average Balancing* are two simple load balancing schemes. In the Global Average Balancing scheme, the cost associated with an AP or BS is changed depending on the deviation of its load from the average load across all APs and BSs in the system. In the case of Local Average Balancing, the average load is calculated only in the vicinity of the AP or BS whose cost is to be updated.

Simulation studies have shown that dynamic pricing based load balancing leads to a higher Packet Delivery Ratio (PDR) for the network. Initially all BSs and APs charge the same access cost of say 5 units. Local average load balancing is periodically applied at the end of windows of 30 seconds duration. Figure 16 shows the marked jump in PDR at time window 2 where balancing is first applied. Thereafter, the PDR shows little variation as load balancing is in effect. Figure 17 shows that the load at the APs surrounding a particular BS has increased after load balancing is applied. Load balancing has resulted in all APs within the particular cell to share a part of the load, which was initially almost wholly concentrated at the BS. The percentage of the total load in the cell handled by the BS reduced from over 98% to approximately 85% after load balancing. Due to the great difference in coverage areas, the load on the BS still remains much higher than that of any of the surrounding APs. Figure 18 shows the change in access cost charged by the APs and BSs after the application of load balancing. We can see that the BS has the highest access cost due to the higher load encountered by it.

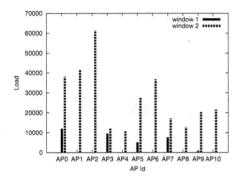

Figure 17: Impact of local load balancing on the load at APs surrounding a particular BS

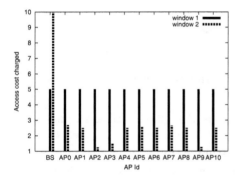

Figure 18: Impact of local load balancing on the access cost charged by APs surrounding a particular BS

5. IMPLEMENTATION ISSUES

Implementing smooth interoperability between wireless LANs and wide area cellular networks is a challenging task. It calls for modifications in the network stacks of the mobile devices' operating systems as well as requires additional features in the WLAN and the cellular network protocols. The applications running on the mobile node may or may not be aware of the dynamic switching between the various network interfaces taking place at the lower layers of the network stack. Switching decisions are based on the mobile node's *User Profile* and the current network conditions. This also calls for devising methods to characterize and measure the current network conditions. The implementation of interoperability in the mobile devices may be done at multiple levels of the network stack – application, transport, network layers. Each layer has its advantages and disadvantages in terms of ease of implementation, speed, efficiency, customizability and extendability.

5.1 Modifications to Support User Profile Based Interoperability in WLANs and Cellular Networks

Almost all currently deployed WLANs are based on IEEE 802.11. Existing wide area cellular networks use GSM or CDMA technologies. All these technologies consist of elaborate beaconing and signaling mechanisms to support the transmission of voice and data traffic over the networks. However as the protocols were never designed with interoperability in mind, they lack certain features which are essential to support user profile based interoperability. Some of the important pieces of information required for interoperability are discussed below.

5.1.1 Current Available Bandwidth

User profile based switching requires knowledge of the throughput attainable in the different networks at any instant. The current network load, the total bandwidth available at each BS or AP, the bandwidth partitioning algorithm, which are required to calculate the current available bandwidth, are all details internal to a particular network and are not available to a user of the network. Periodic beacons containing the free bandwidth information should be broadcasted by the BSs (in CDMA, GSM) and the Access Points (802.11). The dynamic network switching modules in the mobile devices' network stacks extract the bandwidth information from these beacons and make the appropriate switching decisions.

5.1.2 Access Cost Information

To support the *Cost Conscious User Profile*, the network switching module needs to know the access charges associated with the different networks. When the cost is constant across all BSs and APs of a particular network, the fixed value may be embedded as it is into the switching module. In the case of dynamic pricing – *i.e.,* the access cost associated with an BS/AP changes with geographic location, time, network load, etc – the access cost information needs to be conveyed to the mobile nodes through periodically broadcasted beacons, as in the case of free bandwidth information mentioned in the previous section.

5.1.3 Network Coverage Information

The dynamic switching module uses knowledge about the coverage areas of different BSs/APs while making switching decisions for the *Glitch Conscious* users. As in the case of variable access costs, variable coverage areas may be advertised through periodic beacons.

5.2 Design of a Dynamic Network Switching Module

In this section, we describe the design and implementation of a user profile based dynamic network switching module. The function of this module is to take a decision on which network interface is to be used currently. The module reads in the user profile data from its configuration files and periodically queries the various network interfaces to obtain their current status. The structure of the module is shown in Figure 19. This module can be plugged in at multiple layers of the operating system's network stack. The decision to use a particular interface is made by the module, but the mechanism to implement switching will be different in different layers.

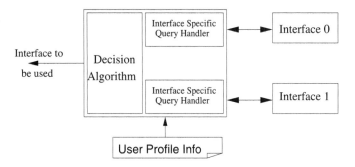

Figure 19: Structure of the User Profile based Dynamic Network Switching Module

5.3 Incorporating Interoperability into the Network Protocol Stack

Interoperability can be incorporated at multiple levels in the network protocol stack of the mobile device's operating system, as described below.

5.3.1 MAC Layer

A-MAC, proposed as part of the AdaptNet protocol suite [9], is a two-layered MAC which can support interoperability. The master sublayer of A-MAC forwards data to the multiple network interfaces on the basis of the *virtual cube* concept. The virtual cube based decision making module can be replaced or supplemented by the dynamic switching module described in Section 5.2 to support user profile based interoperability.

5.3.2 Network Layer

Maintaining a single static IP address for the mobile device across all network interfaces is the main challenge in implementing interoperability at the network layer. A single static IP address for a mobile device can be achieved through Mobile IP [10]. The mobile device is considered to have moved away to a foreign network whenever it switches to a new network interface. This solution requires the Wi-Fi APs and the cellular BSs to support the Home Agents and Foreign Agents of the Mobile IP protocol.

5.3.3 Transport Layer

Implementation of interoperability at the transport layer involves the major challenge of maintaining TCP connections across switching between multiple interfaces. A TCP connection is characterized by the 4-tuple *(Source IP, Source Port, Destination IP, Destination Port)*. As different interfaces have different IP addresses, switching the network interface will break the TCP connection. Redirectable Sockets, *RedSocks*, introduced in [11] is a possible solution to this problem. *pTCP* [12], which achieves bandwidth aggregation by striping data across multiple TCP connections, is another protocol by which smooth interoperability can be attained at the transport layer.

5.3.4 Application Layer

Applications may be written such that they are aware of the multiple network interfaces, and may incorporate application specific considerations into the switching decisions. All calls to open a socket in the application must be modified so as to attach the socket to the interface specified by the switching module rather than to the default interface. In Linux, this involves appropriately setting the SO_BINDTO-DEVICE socket option. While application specific optimizations are possible, the disadvantage of implementing interoperability at the application layer is that all applications using interoperability are to be modified or rewritten from scratch.

5.4 *interoproxy* - **An Application Layer Implementation**

interoproxy, is an application layer implementation of interoperability. This interoperable HTTP proxy uses the dynamic network switching module for choosing the interface over which the HTTP requests are to be made, on the basis of the user profile and the current network conditions. An innovative HTTP request replay mechanism is also added to the proxy to take care of requests lost during switching from one interface to another. We must note here that *interoproxy* runs locally on the same machine as the Internet browser and is used even when a direct connection to the Internet is available.

interoproxy was implemented by extending *tinyproxy*[13], a lightweight HTTP proxy licensed under the GNU Public License (GPL), to support user profile based dynamic switching between multiple network interfaces. Modifications to the source code were limited to only the places where sockets were opened to relay the web browser's request to the Internet or to an upstream proxy. The tinyproxy opensock function was modified to bind the socket to the interface specified by the interface selection logic using the Linux SO_BINDTODEVICE socket option.

5.5 Smooth Interoperability through Request Replay

Smooth interoperability requires that the user remains unaware of the dynamic switching taking place between the multiple network interfaces. Consider the following situation. A user requests a web page through his web browser. *interoproxy* opens a TCP connection to the web server over the CDMA interface. Before the HTTP reply was received, the CDMA interface got disconnected and the proxy switched to the WLAN interface. In such a case, the proxy will automatically replay the requests sent out on the broken interface for which no reply was received, on the new interface. Thus the user is unaware of the switching and gets the desired web page. In some cases, the switching may take place even though the current interface is not completely disconnected. In such a case, the proxy may decide to wait for replies to already placed requests on the old interface or it may choose to terminate the active HTTP sessions and replay the requests on the newly active interface.

6. CONCLUSION

Interoperability between wireless LANs and wide area cellular networks is a very challenging task. Mobile devices of the future, equipped with multiple network interfaces, will dynamically switch between the interfaces on the basis of their user profile.

Simulation studies have shown that user profile based switching leads to higher network performance and increased user satisfaction. Pricing and revenue sharing are extremely important in an interoperable scenario as the commercial viability of all the service providers involved must be assured. Various pricing schemes and revenue sharing schemes were analyzed in this paper. It was also shown that load balancing between the cellular BSs and Wi-Fi APs on the basis of dynamic pricing leads to an improvement in the overall network performance.

7. REFERENCES

[1] GSM World, http://www.gsmworld.com.

[2] CDMA Development Group, http://www.cdg.org.

[3] V. Sekar, B. S. Manoj, and C. Siva Ram Murthy, "Routing for a Single Interface MCN Architecture and Pricing Schemes for Data Traffic in Multihop Cellular Networks" in *Proc. IEEE ICC 2003*, pp. 969-973, May 2003.

[4] Y. D. Lin and Y. C. Hsu, "Multihop Cellular: A New Architecture for Wireless Communications," in *Proc. IEEE INFOCOM 2000*, pp. 1273-1282, March 2000.

[5] R Ananthapadmanabha, B. S. Manoj, and C. Siva Ram Murthy, "Multihop Cellular Networks: The Architecture and Routing Protocols", in *Proc. IEEE PIMRC 2001*, pp. 78-83, October 2001.

[6] D. B. Johnson and D. A. Maltz, "Dynamic Source Routing in Ad hoc Wireless Networks", *Mobile Computing, Kluwer Academic Publishers*, vol. 353, pp. 153-181, 1996.

[7] Dilip Antony Joseph, B.S. Manoj, and C. Siva Ram Murthy, "The Interoperability of Wi-Fi Hotspots and Packet Cellular Networks and the Impact of User Behaviour", to appear in *Proc. IEEE PIMRC 2004*, September 2004.

[8] X. Zeng, R. Bagrodia, and M. Gerla, "GloMoSim: A Library for Parallel Simulation of Large-scale Wireless Networks," in *Proc. PADS-98*, Banff, Canada, May 1998.

[9] I. Akyildiz, Y. Altunbasak, F. Fekri, and R. Sivakumar, "AdaptNet: An Adaptive Protocol Suite for the Next-Generation Wireless Internet", IEEE Communications Magazine, vol. 42, no. 3, pp. 128-136, March 2004.

[10] C. Perkins, "Mobile IP", IEEE Communications Magazine, vol. 35, no. 5, pp. 84-86, March 1997.

[11] M. Haungs, R. Pandey, E. Barr, J. F. Barnes, "A Fast Connection-Time Redirection Mechanism for Internet Application Scalability", in *Proc. HiPC 2002*, pp. 209-218, December 2002.

[12] H. Y. Hsieh and R. Sivakumar, "A Transport Layer Approach for Achieving Aggregate Bandwidths on Multi-homed Mobile Hosts". in *Proc. ACM MOBICOM 2002*, pp. 83-94, September 2002.

[13] tinyproxy - A Lightweight HTTP proxy, http://tinyproxy.sourceforge.net

A User-Centric Analysis of Vertical Handovers

Andrea Calvagna, Giuseppe Di Modica
Dipartimento di Ingegneria Informatica e delle Telecomunicazioni
Università di Catania
Viale A. Doria 6, 95125 Catania Italy
Email: Andrea.Calvagna@unict.it,Giuseppe.DiModica@diit.unict.it

ABSTRACT

To implement seamless mobility inside an integrated, multiple (e.g. GPRS/WiFi) access system, a vertical handover policy has to be devised. This is usually done at the mobile terminal, allowing it to be customized from a end-user perspective, in order to fit individual needs/preferences. We propose a new approach in taking vertical handover decisions, which are not anymore exclusively based on the knowledge of the available access networks' characteristics but also on higher level parameters which fall in the transport and application layers. To this extent, in this paper a model has been realized and simulations have been run in order to evaluate the impact of the vertical handover and its frequency on a set of typical user's network applications/services. We also take into account the user preferences in terms of cost and quality of service. We believe this approach reflects the optimal settings from the user's point of view with regard to his running services and applications. Our aim is to understand how to define a metric to be used in order to devise a solution which should try to balance the overall *cost* of vertical handovers with the actual benefits they bring to actual user's networking needs. This way, each mobile user could autonomously apply the handover decision policy which is more convenient to his specific needs.

Categories and Subject Descriptors

C.2.1 [**Computer-Communication Networks**]: Network Architecture & Design—Wireless Communication

General Terms

Design, Performance

Keywords

seamless mobility, handover, user satisfaction, TCP, UDP

1. INTRODUCTION

The 4th Generation (4G) networks concept support wideband data and telecom services for mobile users roaming across multiple, wireless and wired, integrated access networks. Basically, the purpose is that of combining all the existing heterogeneous wireless networks into a single, interoperable system, being IP protocol the "glue" between the set of underlying radio access and physical layers. Within this overall scenario, it is foreseeable that many access technologies, even with very diverse profiles (in terms of bandwidth, latency, security, etc.) will often be available in overlapping areas. Users will also be equipped with terminals capable of multiple access interfaces, or provided with a dynamically reconfigurable access interface [13], allowing them to seamlessly take advantage of more than just one physical access connection, even at the same time. Particularly, in the context of overlapping, heterogeneous, access networks, specific strategies must be devised to control the triggering of "vertical" handovers [20], between the available access networks, which affect the overall performance of application sessions running on the mobile user's device. Also, fast handover procedures are needed to meet the applications requirements. In fact, the handover latency may heavily affect the continuity of application sessions (let us think about the strong delay requirements imposed by real-time applications).

In this context, current literature is focused on optimizing vertical handovers from the network-level point of view only [14, 6]. In contrast, the specific design of vertical handover policy may deeply impact on the performance of user applications at the user terminal. As an example, a triggering policy which simply switches to any "better" (e.g., lower latency) access network as soon as available, could disappoint the user with possibly frequent connection discontinuities and, depending on the running applications (e.g. non real-time) would not necessarily improve the performance. Other non network-level parameters and variables, such as the available connections cost, the user's mobility pattern, the kind of the applications running on the mobile terminal, should be accounted for when considering what is "better" in overall from the end user point of view. In this paper we modeled a possible 4G network scenario and performed some simulation work to assess the impact of vertical handovers on the main kinds of IP-based application flows.

The paper is structured as follows. In Section 2 the related works and the motivation of our work are discussed. In section 3 the concept of "cost" of vertical handovers from the the point of view of the transport and the application layers is introduced. Section 4 presents the proposal of a user-centric middleware architecture. In section 5 a model of interworking between WLAN and GPRS is described, in which simulations have been carried out, whose results are presented in section 6. In section 7 the implementation of a real wireless mobility framework is described. Finally, section 8 concludes our work.

2. RELATED WORKS AND MOTIVATION

Several schemes have been proposed to handle vertical handovers between heterogeneous wireless networks. When designing a vertical handover scheme, two aspects are to be taken into account. On one hand, the handover should be as smooth as possible: when being handed over, the MH should only experience a change in the perceived bandwidth and end-to-end packet delay. On the other hand, an decision algorithm is needed to schedule the time for the handovers, according to some QoS criteria.

In the BARWAN project [20] the performances of handover are improved by employing multicast and buffering techniques, but the requirements imposed by this scheme currently seems to prevent any feasible deployment. The solution proposed in HOPOVER [6] relies on a resources early-reservation scheme and on buffering techniques as well, claiming to solve the problems incurred by the BARWAN's solution. Buffering techniques are widely employed by all schemes that aim at guaranteeing smooth and seamless handovers. In [19] authors have investigated the upward vertical handover in wireless overlay networks, and have determined buffer requirements for lossless handovers. In [17] a seamless vertical handover procedure is presented, together with an effective algorithm for handover decisions based on the requirements imposed by the running application sessions whose continuity is to be granted. In [14] authors propose another optimization scheme for mobile users performing vertical handovers. The profitability of a vertical handover is stated according to performance parameters such as mean throughput and handover delay.

As far as vertical handover decision algorithms are concerned, all the schemes proposed in literature, as well as the one we use, borrow from techniques widely employed in the wireless cellular networks. Of particular interest are the ones based on the averaging window and hysteresis margin [15], those using pattern recognitions [25], and the fuzzy-based [7, 16, 22]. The common target of these techniques is to minimize the so called "unnecessary handovers" while maintaining the QoS constraint. In wireless cellular network, minimizing the number of handovers is important as each handover increases the signal and processing load, and causes traffic management problems, thus affecting the QoS. In general, for such networks the QoS constraint related to the mobility management is the handover failure probability, also referred to the forced termination probability of handover calls.

However, when trying to apply these techniques to heterogeneous environments, where several networks with different wireless access technologies are put together to transport data, the concept of QoS is to be reviewed. IP connectivity maintenance of mobile users roaming though heterogeneous wireless networks seems to be the main concern of the vertical handover algorithms. The focus is at network level, being IP the umbrella for the several different wireless access protocols. This work tries to shift the focus from the network layer to the transport layer and even to the application layer. Handing over the mobile user from one network to another impacts on his ongoing application sessions. Of course, techniques can be devised to make the vertical handover as seamless as possible. But it is a matter of fact that a vertical handover to/from a different wireless network causes a change in the connection parameters as perceived by the mobile user (e.g., in terms of throughput and end-to-end RTT). On long term, several continuous ping-pongs among heterogeneous networks, and therefore frequent fluctuations of the connection parameters, affect the performances of the application sessions running on the mobile user's device.

This work proposes an analysis of the impact that such fluctuations have on the performances of main Internet applications. Schemes have been proposed to speed up the performances when vertical handovers are triggered [11, 3, 12, 9, 2]. We do not introduce new techniques for a smooth vertical handover, neither propose a new handover decision algorithm. Instead, we report an analysis of how the tuning of the hysteresis margin of a very straightforward handover decision's algorithm influences the performances of applications' sessions.

3. THE *COST* OF VERTICAL HANDOVERS (VERTICAL HANDOVER POLICIES)

The overall, network-level cost of performing a vertical handover between two types of access systems could be expressed in terms of the latency and bandwidth gaps between the two. Also, the location update latency imposed by the installed IP mobility manager (e.g. MIP [18]) has to be considered as an additional delay. What we now have asked instead is: how can we measure how this vertical handover will influence the behavior of running applications? In other words, what is the application-level cost of the very same *network layer* vertical handover, which may be very diverse depending on how it will impact on the application and the protocol it uses to manage end-to-end data flows (UDP, TCP, RTP, HTTP, FTP, etc.).

Thus, we shift the focus from the network layer to the transport and above layers. Of course, techniques can be devised to make vertical handovers as seamless as possible to this point of view. It is a matter of fact that a vertical handover to/from a different wireless network causes changes in the connection throughput and end-to-end RTT (*round trip time*), and frequent fluctuations of the connection parameters may affect negatively the performances of the application sessions running on the mobile user's device.

As a consequence, it may be argued that the final effectiveness of a vertical handover policy should have to be valuated over the whole length of a mobile user's work session, and from a higher-level point of view. Obviously, the ideally optimal schedule of vertical handovers cannot be exactly computed unless the running applications, the access networks topology and the user's movement pattern inside it are all well known beforehand. By the way, it is possible to use abstract modeling to derive a more generally applicable, near optimal, solution.

4. AN ARCHITECTURE TO SUPPORT THE USER PREFERENCES

Providing 4G mobile-users with the capability to seamlessly and conveniently keep their active work sessions across a network of multiple, possibly overlapping, heterogeneous radio access technologies, requires as a fundamental step to define what "best connected" means at the user-level, i.e. from their personal point of view. We argue that this user-level definition of what is "best" in such context may not be universally and statically determined. In contrast, it is a dynamic concept strictly related to many factors including:

- the attributes characterizing the current set of applications the mobile user is running in his mobile-terminal (let's call it the "running applications profile");

- the attributes characterizing the access conditions offered by the set of locally available radio networks (the "available access profiles");

- the overall network topology;

- the user's movements pattern

Despite all of the above mentioned (except for the last one) are somewhat network-resource usage related attributes, the user-level "settings" may reflect also strong implications outside the purely functional domain, i.e. the cost. In fact, private companies are willing to profit from their offer of 4G access networks and services, which will thus have significant fees associated to their usage. As a consequence, money should be one of the key parameter driving user-level optimization of commercial networks' resource usage. To summarize, we argue that a 4G mobile user should be able to use his terminal to dynamically evaluate locally offered network services (access/data) and optimize functional (performance of running applications) and non functional (cost of services) requirements according to his specific needs. In this section we propose a possible middleware architecture for the deployment of an ABC system [8, 10]. In Figure 1 a picture of the proposed architecture is depicted. In the following, a description of each functional block is given, with particular emphasis given to the Profile Management Module.

4.1 Wireless access network detection and monitoring.

The detection and the monitoring of the wireless access networks visited by the end user while roaming is one of the most challenging problem to be faced. The *network monitoring module* is in charge of interfacing to all the network cards that the user terminal is supplied with, and continuously gathering information about the availability and the reliability of the different wireless access channels. A mechanism is needed to collect the heterogeneous information from the different networks and uniform them (for instance, by means of normalizing procedures). Furthermore, other network relevant data can be collected at this stage. The network operator might want to provide some useful information concerning its brand, the actual bandwidth occupancy, the level of network QoS that can be supplied at that moment to that user (e.g., in terms of bandwidth availability and packet delay), the coverage area granted by that specific wireless access point.

4.2 User profile management.

One of the main ideas that drove us throughout the design of the middleware is giving the end user as much control as possible on the selection of the wireless network that best

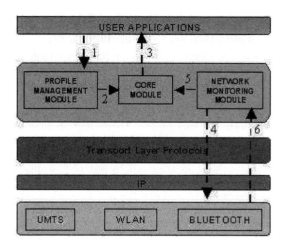

Figure 1: (1) Interaction between the User and his User Profile. (2) The user preferences as an input to the network selection process. (3) The feedback to the user applications. (4) The selection of the wireless network. (5) The data from the networks as an input to the network selection process. (6) The monitoring of the available networks

meets his preferences. In our opinion, user satisfaction [24] is to be taken in great account when trying to enable the concept of "always best connectivity". The perception of "best" connectivity changes from user to user, and strictly depends on the relative value that the user gives to one aspect of the connectivity with respect to another. The *user profile management* module gives the user the chance to statically specify his own preferences, and dynamically modify them whenever he needs. The preferences are locally stored (i.e., in the terminal device's memory, or in the USIM), and its management does not involve any remote interaction. For instance, the user might wish to 1)save on the connection cost, by preferring, when available, the networks whose connection costs are lower, according to the fares of the contracts that he has subscribed to; unavoidably, this strategy would reduce the number of accessible networks, and definitely no guarantee is given about the sessions' continuity (application sessions might be abruptly broken down because no connection is available according to the user preferences); 2) guarantee his ongoing application sessions as much as possible, no matter the connection costs; 3) find some compromise between cost savings and sessions' continuity; 4) specify the set of applications that must be preserved in the case that the network performances degrade (e.g., because of an handover); 5) specify whether some applications must be alerted whenever the network conditions change. The *network selection process* is given access to these preferences that, together with the data collected from the network, are used to make the "best" network choice. It is worth noting that a network selection (i.e., an handover to a different network) can be performed not only because of the sudden unavailability of the currently accessed network, but also because the latter no more meets the actual preferences expressed by the user.

4.3 Wireless access network selection and application session management.

The network selection process module is in charge of continuously monitoring the networks. To this end, data revealing the actual status-condition of the available networks are periodically retrieved form the network detection module. The *core module* will combine such data with the ones retrieved form the user profile. Depending on whether the QoS expressed by the user in his preferences can be sustained by the current network conditions, a change of the current network can be triggered, by selecting the next from the available ones, or some applications might be alerted to let them accordingly adapt to the new network conditions. In the first case, the new network will be selected according to the user's need (as specified in the user profile), and in particular the one that guarantees a reliable connection support will be chosen (according to the data gathered from the each network interface). In the second case, a feed back of the changed network conditions is provided to some applications that the user wish to preserve. On their turn, the applications will try to adjust their parameters, so that a minimum functionality is granted as specified by the user. Let us focus on the following scenario. Suppose that the user is accessing a network providing him with a good connection in term of available bandwidth. He decides to run an adaptable video stream session, and sets the parameters of the service accordingly to the high bandwidth that he is given (e.g., high frame and bit rate, full color, etc...). Suddenly the current network conditions vary in such a way that the current session can no more be sustained. According to the user preferences, two different solutions are possible: (1) the client-side video stream application is alerted about the changed network conditions, so that a downgrading in the application parameters is imposed (e.g., the frame and bit rate might be lowered, the audio suppressed, etc..); (2) a new network is looked up, able to sustain the current application session, either in full mode (high frame and bit rate) or in half mode (low frame and bit rate), depending, once again, on the user convenience.

5. MODEL DESCRIPTION

We have developed a model implementing a subset of the requirements for the interworking between WLAN and GPRS. A MIP-like distributed mobility protocol has been designed and integrated in the model to support the roaming of mobile hosts (MHs) in the WLAN and in the GPRS domain. A Cellular IP (CIP) [5] derived protocol is used to take care of micro-movements management. In particular, an hysteresis-based strategy has been conceived for horizontal handovers triggering.

According to [20], we refer to a WiFi-to-GPRS handover as an upward vertical handover; such an handover is in general triggered when the MH moves out of reach from a narrow-coverage (but broadband) network while already inside an overlaying wide-coverage (and narrowband) network. The common upward vertical handover strategy is very straightforward. Beacon packets from every available WiFi Access Points (APs) are constantly monitored, with particular attention to the ones collected by the AP that is currently serving the MH. Whenever the signal strength level of these beacons falls below a given threshold, meaning that the current radio signal is going to be lost, a new AP is

chosen, among the monitored ones, in order to hand over to it all MH's ongoing sessions (in this case, an horizontal handover would be triggered). If none is available, or their radio signal strength is too weak, an upward vertical handover is triggered to the GPRS network. The MH's data connections might be heavily affected by this kind of vertical handover, since (by definition) it is just the consequence of a lack of connectivity (radio "silence") experienced by the MH, whose handling over to the wide-coverage network is not immediate but requires also another, so called, handover latency. Conversely, we refer to a downward vertical handover when the MH leaves a wide-coverage network to move to a narrower one (GPRS-to-WiFi handover). Triggering of this kind of vertical handover is usually not a need, as was the previous case, but just an opportunity. As a consequence, it is less disruptive than the previous, given that the MH, during the handover phase, can keep its ongoing connections alive in the overlaying network until the handover procedure has been completed. Let us now assume that the MH is currently accessing the GPRS network, and that the MH is approaching a region covered by a WiFi AP. In our model, the following threshold-based strategy is adopted to schedule the time for a downward vertical handover. As soon as the MH senses the first beacon from the WiFi network card that it is equipped with, a timer is started. The average signal strength of the received beacons (ARSS) is monitored until the timer expires. At expiration time the signal's ARSS is checked against a given value Th, representing the strength level that the ARSS at least has to equal in order for the signal to be assessed reliable. The greater Th, the harder for a signal to be evaluated reliable. The purpose of this scheme is to filter out all the weak and intermittent signals coming from nearby WiFi APs, thus avoiding unnecessary downward vertical handovers that might result in further heavy sessions' disruptions. In fact, if a downward vertical handover were performed to a WiFi AP, whose fading radio signal soon revealed too weak to support the MH connection, a new upward vertical handover should immediately be triggered to divert the MH's communication sessions back to the GPRS network. In such scenario, given the higher handover latency time needed to complete an upward vertical handover, the well know ping-pong effect is even more disruptive than in scenarios where only horizontal handovers occur.

Simulations have been carried out in a NS2 [23] model that we have developed. The MH moves within a 700x550 points rectangular area according to the well known mobility scheme named "random walking model"; the pattern of the MH's movements have been generated being the parameters max-speed set to 20 m/sec and the pause time set to 0 sec. Several 802.11 APs have been placed in such a way to ensure islands covered by radio signal to the roaming MH. The transmission power of the APs has been set up to a value that guarantees overlapping areas of radio signals. Furthermore, regions not covered by any radio signal ("holes") have been intentionally left in between radio covered islands. In such holes the MH will take profit of the overlaying GPRS network.

6. SIMULATION RESULTS

6.1 Used traffic models

In [21] a classification of the traffic running over the Internet is given. Elastic traffic includes TCB-based applications like Telnet, FTP, P2P file sharing, Email and Web browsing. Even though reliability is a crucial QoS parameter for these application, throughput may be considered a performance metric as well. Inelastic traffic is generated by real time services (voice and video) and, in general, by all the data services to which both timing and throughput are relevant parameters in order to meet the QoS requirements. As far as the elastic traffic is concerned, a further classification can be introduced, based upon the level of user interactivity imposed by the semantics of each application. The transfer of a long-sized file, as well as the downloading of an email with a big attachment, generates long-lived TCP transmissions with no interaction from the user. Conversely, applications like Web browsing and Telnet envisage a tight user interaction but generate a lot of short-lived TCP transmissions. The inelastic traffic category enumerates applications like VoIP, MPEG and H.263 video sources. Given the strict timing requirements, they reside on top of the UDP transport layer and usually generate transmissions at constant or variable bit rate. In the proposed analysis we make an effort to cover all the categories of Internet traffic, by employing several traffic source models.

6.2 Performance and metrics

Let Sm be the minimum signal strength level in order for a packet to be correctly sensed by the MH when it roams in a WiFi domain. For each category of simulation, several simulations will be run, respectively setting Th to different values. In particular, one simulation will be run by setting the threshold to such a value that the MH will never abandon the GPRS connection (let us call this conservative settings "connection-safe"). Another simulation is run by instructing the handover algorithm to search for just WiFi access points, even if that means that the MH will experience "black-outs" in its connections (let us call this settings "bandwidth greedy"). Other simulations are run by setting the Th to values $1.1*Sm$, $1.4*Sm$ and $10*Sm$ respectively. A complete set of simulations is thus run, ranging from the most connection-conservative configuration to the most bandwidth-greedy. Results will show how varying the Th affects in different ways the performance of the running applications. We can monitor the performance of the TCP-based applications by observing how fast the number sequencing of the TCP segments increases and/or how big is the amount of transferred packets. At the end of a simulation, the greater the sequence number, the better the performance of the TCP. As far as the UDP-based applications are concerned, the relevant metrics that we will monitor are the total number of lost packets and the packet delay as a function of the time. In order for UPD to show good performance, both the packet loss and the fluctuation's rate of the end-to-end packet delay are to be kept as low as possible.

6.3 TCP-based simulations

6.3.1 FTP session

The first group of simulations is focused on the study of the dynamics of the TCP protocol, during vertical handovers, when a file transfer is going on both upwards (from WiFi to GPRS) and downwards (GPRS to WiFi) between the MH and a corresponding host(CH) in the Internet. The FTP session is set up to start at time t=1.0s, and lasts until the end of the simulation: it is a typical, non-interactive, long-lived TCP transmission. In Figure 2 the sequence number progression of the TCP segments received by the CH is plotted for different configurations of handover threshold. The simulation that performs better is the one with $Th=1.1*Sm$. We notice that its performance does not differ much from that of the "bandwidth-greedy" (BG) simulation setting. Even if the latter experiences long "black-out" periods, the TCP protocol is able to greatly evolve whenever the WiFi access is available (thanks to the higher bandwidth and lower RTT). Conversely, the simulation with "connection-safe" (CS) settings does not experience connection blackouts, but can only benefit from a connection with limited bandwidth and a high RTT value (the GPRS one). We can conclude that for long-lived TCP transmissions (like, for instance, a file transfer), "bandwidth-greedy" handover strategies seems to perform better. For this kind of transmissions it is more advisable a connection with high bandwidth and low RTT, even if intermittent, rather than a permanent but poor connection.

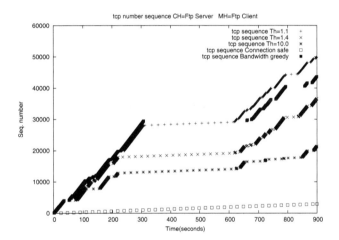

Figure 2: The TCP number sequencing

6.3.2 Http session

In this set of simulations an Http session is started between the MH (the http client) and the CH (the http server). The client's page request generation process follows the Pareto distribution model. This model has been used in order to simulate the typical behavior of the user that browses the Web: a new page request is issued after the relevant information contained in the just downloaded page have been read by the user. The average time spent by the user to read a page (i.e., the average $Toff$ in the Pareto distribution) is set to 15 seconds, while the average web page size (i.e., the one that characterizes the Ton in the Pareto distribution) is set to 8 Kb. Obviously, the resulting Ton is variable, since depends on the characteristics of the network currently accessed by the user. The traffic model that is being simulated is short-lived, and a loose user interaction is observed. In Figure 3 the total amount of the downloaded traffic for each simulation is shown. Once again, the best performance is obtained by setting $Th=1.1*Sm$. This time, the simulation with BG settings shows the worst performance, while the one with CS settings is not greatly penalized. The Http traffic model, in fact, gives rise to short-lived TCP connections. The application does not greatly benefit from the higher bandwidth available in the WiFi domain, given that the most of time is spent by the user to read the downloaded page, whilst the connection is exploited for a very little fraction of time. That is why the simulation in which the MH has a permanent connection to the GPRS network performs better than the simulation in which the MH connects only to the WiFi network. In fact, in the latter, from time t=270s to time t=620s the MH's connection gets stuck because of the WiFi black-outs (i.e., the MH is out of the range of any WiFi access point), while the former can benefit from the GPRS connection. The conclusion that we draw is that the connection parameters (the bandwidth and the RTT) have a lower influence on the overall performance when short-lived TCP transmission are considered.

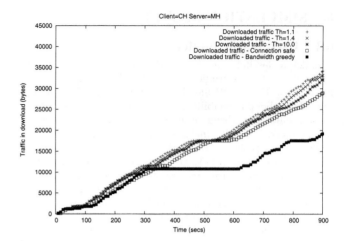

Figure 4: Downloaded traffic during the telnet session

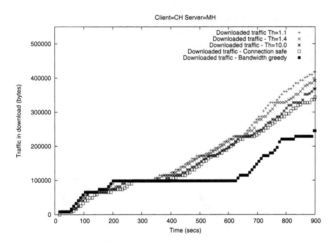

Figure 3: Downloaded traffic during the http session

6.3.3 Telnet session

We describe the results that we obtained by simulating a telnet session. This kind of application gives rise to a traffic model that resembles the one generated by a Http session. The user, in fact, interacts with his terminal, but this kind of interaction is tighter than the one observed for the Web browsing session. Furthermore, the size of the packets exchanged between the telnet client program and the server one are much smaller than those exchanged in http sessions. We refer to this kind of traffic as a very short lived one with high user interaction. In Figure 4 the measured performance for each simulation is reported. All the simulations, except the one with BG settings, seems to show equal performance. Once again, the strategy of searching at any cost for the WiFi access does not pay.

6.3.4 UDP-based simulations

Simulations have been run to evaluate the impact of vertical handovers on the performances of applications built on top of the UDP protocol. In particular, RTP sessions have been set up between the MH and the CH. A CBR application running on the CH sends packet at a rate sustainable by the capacity of the link that has the narrower bandwidth (i.e.,

the GPRS link). This is to make sure that no packet gets lost due to congestion in any queue of the traversed domains' routers. In Figure 5 the total number of lost packets for each simulation are shown. The graph refers to the number of packets that have been lost during handovers, both horizontal and vertical. The more the threshold increases, the less packets are lost. Of course, the simulation with CS settings gives the best result (no packet lost), while the one with BG settings (not shown in the graph) experiences a great packet loss (more than 8000 packets get lost). In Figure 7(a),7(b),7(c) the measured end-to-end packet delay is plotted for the simulations with $Th=1.1*Sm$, $Th=10.0*Sm$ and the one with bandwidth-greedy settings have been reported. During the 900 seconds of simulations, the MH switches several times from the WiFi network to the GPRS. The ongoing CBR session undergoes frequent packet delay fluctuations (from 0.018 seconds in the WiFi link to 0.55 in the GPRS link) as far as the experienced throughput and the RTT are concerned. In particular, for this specific simulation, only the end-to-end RTT is affected by the frequent vertical handovers, given that the transmission rate of the CBR session does not exceed the GPRS link capacity. The graphs show that the rate of the packet delay fluctuations lowers as soon as the threshold increases. As far as the BG simulation is concerned, no fluctuations can be found, since the packet delay is almost constantly set to 0.018 seconds; but of course a lot of packets are lost during the black-out periods. The results of the simulation with CS settings (whose graph is not reported) showed a constant packet delay almost equal to half the RTT of the GPRS link.

Let us define *inter-handover duration*, or *interval*, as the interval spent by the user within a wireless access network before the next vertical handover occurs. In Figure 6 by varying the duration of inter-handover times we report respectively the number of occurrences of such intervals with respect to their amplitude in seconds. It can be noticed that by comparing the simulations, the number of times that the *inter-handover interval* is about ten seconds respectively decreases from 18 for $Th=1.1$, to 13 for $Th=1.4$, to 4 for $Th=10$. Generally, when varying the Th from 1.1 to 1.4 and 10, the occurrences of short *inter-handover intervals* tend to diminish while those of longer ones increase.

In the graph, the lower-right points showing durations of respectively 300, 393 and 400 seconds represent the long, continuous, permanence of the MH in the GPRS network respectively for each of the three simulations (see also Figure 7(a) and 7(b)).

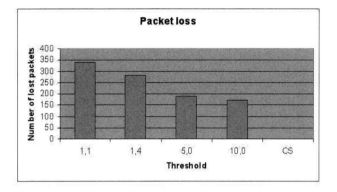

Figure 5: Total number of lost packets

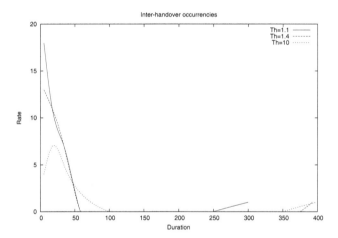

Figure 6: Occurrences of inter-handover times

6.3.5 *Overall connection* cost

The previous simulation have been carried out with the aim of evaluating the impact of vertical handovers just on the performance of the applications running on the user terminal. The results have shown that by varying the handover decision policy the performance of some applications improves while that of others might worsen. According to the application actually running on the user terminal, one could try to find the optimum policy (i.e., the optimum value for the threshold Th) to maximize its performance. Obviously, this strategy would perform the scheduling of the wireless network, disregarding non functional parameters like, for instance, the connection cost charged by the network provider to the user.

While roaming, the user is offered several wireless access by different network providers. The fees he will have to pay depends on the type of contract that he has subscribed, on the roaming agreements between the providers whose networks he is going to access, and of course on the actual usage of the offered connections. The user will likely be charged based upon the time he spends in a given connection or upon the amount of data transferred through the connection itself. In Figure 8 we report the time spent by the user in the WiFi and in the GPRS networks, respectively for each of the policies that controls the handovers. If we assume that in this specific case the user is not charged for using the WiFi connection, and therefore the overall due fee is proportional to the amount of time that he spends in the GPRS network, a suitable handover policy can be chosen according to his willingness to pay.

In particular, for a given communication session, let us define the following *cost* function:

$$C = \sum_{i=1}^{n} T_{Ni} \cdot c_{Ni}(h) \tag{1}$$

where T_{Ni} is the sum of the permanence intervals spent by the user in the i-th access network and $c_{Ni}(h)$ is the fee per unit of time (second) that the operator of the i-th access network charges to the user. The 1 represents the monetary cost faced by the user for a given communication session. It is worth noticing that $c_{Ni}(h)$ may vary, depending on the actual time at which the user is accessing the i-th network (think about some fares' plane, according to which the operator charges higher fees to the user if he access the network at specific hours of the day). The term T_{Ni} strictly depends on both the user's movements pattern and the adopted handover decision's policy. When applied to our scenario, the 1 becomes

$$C = T_{WiFi} \cdot c_{WiFi}(h) + T_{GPRS} \cdot c_{GPRS}(h) \tag{2}$$

Thus, given the time that the user is accessing the networks (h) and the costs associated to the usage of each single network ($c_{WiFi}(h)$ and $c_{GPRS}(h)$), by adopting a suitable handover decision's policy (in this case, by tuning the threshold Th) the willingness to pay expressed by the user can be reasonably satisfied.

However, the adoption of a policy that allows the user to save money does not give guarantees that the performance of the running application sessions will be preserved. For instance, an handover strategy with BG settings on one hand might satisfy the user from the connection cost point of view, but will disappoint his expectation of QoS, depending on the application sessions currently going on: this is the case of an http session or a telnet session, whose performance does not benefit from such a policy (see Figure 3 and 4). Conversely, a conservative policy like the CS one will satisfy that user who is willing to pay for having its connections as granted as possible, but might reveal ineffective if, for example, a long file transfer is currently going on (see Figure 2).

That is to say that it is not always possible to find an optimum policy meeting both functional (applications performance) and non-functional (user's willingness to pay) requirements. Most of times, a balance is to be found between the money that the user is willing to pay for his services, and the QoS' degradation that he is willing to accept.

7. REAL SYSTEM IMPLEMENTATION

In this section we describe an actual wireless mobility framework [4] we designed and implemented in our University Campus, which is spread over a wide metropolitan area.

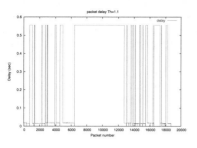
(a) Packet delay when Th=1.1

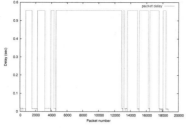
(b) packet delay when Th=10

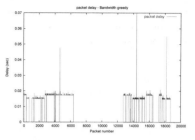
(c) Packet delay for BG settings

Figure 7: Packet Delay

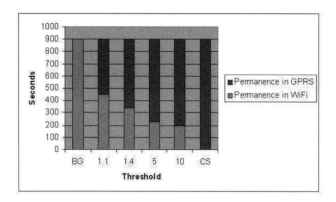

Figure 8: Time spent by the MH in the GPRS and in the WiFi network

The framework is the step that the process of implementation of the user-centric middleware presented in section 3 will start from.

The aim of the framework in its current implementation is to let a mobile host, that is a user equipped with a PDA device, experience real wireless IP mobility while moving on a large spatial scale, by means of a middleware that leverages from three main wireless access technologies: Bluetooth, WiFi and GPRS. In our framework, these are managed as a hierarchy of spatially overlapping access domains. While the user is on the move, the client-side of the middleware running on our mobile PDA device triggers smart switchings between the currently best available and more appropriate wireless access path, based on locally collected information.

We considered as mobile host a PDA device (IPaq 3870) running Linux. Thanks to a WiFi card installed on it, the device is free to move around into a domain offering 802.11b connectivity and mobility support through our mobility middleware, which extends from the Cellular IP (CIP) micromobility protocol, thus experiencing service continuity inside or between spatially contiguous wireless LAN contexts. The PDA we used is capable also of integrated Bluetooth [1] connectivity: this will be a key feature in enabling our mobility framework. In fact, the used PDA device does not expose any integrated GPRS access interface, and PDA expansion units featuring both GSM/GPRS and WiFi radio access interfaces were not yet commercially available at the time of our experimentation. Thus, we used the PDA's inte-

grated Bluetooth interface to create a personal area network (PPP/BNEP) link to a separate Bluetooth-enabled mobile phone,which in turn offered also GSM/GPRS connectivity. To summarize, when the user brings its PDA outside the radio boundaries of a WLAN enabled domain, the device middleware will automatically detect the loss of WiFi connectivity and, as a consequence it will divert all IP connections to the GPRS access network. This will be reached by means of the Bluetooth PAN specifically set-up between the PDA and the GPRS mobile phone. As a result, a user carrying both a PDA with some running Internet applications and a BT+GPRS cellular phone in his pocket can seamlessly experience wireless IP mobility inside our metropolitan campus area. The logical scenario which abstracts the physical context is depicted in Figure 9. Our mobility middleware extended the original CIP middleware to allow the MH leverage from the availability of not just one but two radio access interfaces, 802.11b and GPRS via a Bluetooth link, switching between them when appropriate in order to always stay connected. Specifically, the PDA network connections are carried through the GPRS access domain when WiFi access is not available, but are carried back to the WiFi access domain as soon as an available WiFi access point is detected by the PDA middleware. The whole software architecture in the MH is depicted in Figure 10, showing in dark color the parts that were written from scratch and in light gray those other that only needed to be adapted/modified for the embedded version of the Linux OS. As can be seen, there is a strict correlation between the CIP and the Switcher module, which practically extends its basic functionality (WLAN, intra-domain handover) to allow for a new type of handover between heterogeneous access networks (inter-domain).

Three types of wireless access are considered:

- IEEE 802.11: This module is the wireless IP interface and drive used to establish a data-link layer connection between MH and current BS. It is the only interface actually used by the MH when inside a CIP domain. No substantial modification has been introduced in this module.

- GPRS: The MH uses GPRS to access the network when it is outside a WLAN.The IP packets are encapsulated into point-to-point data-link frames. The other PPP endpoint is the GPRS network interface offered by a mobile-phone. Likely,the MH and GPRS mobile phone are physically connected through a serial

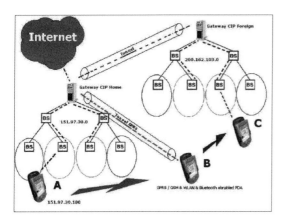

Figure 9: Considered system scenario:MH moves out from its home domain (A)toward a distant foreign domain (C),preserving connections to the Internet through GPRS meanwhile(B)

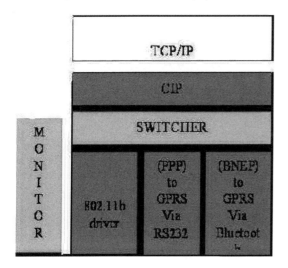

Figure 10: Software Architecture in the MH

RS232 cable.

- Bluetooth: Most of the commercial mobile computing terminals do not have GPRS interface.On the contrary, they are expected to offer a Bluetooth(BT) interface. This can be utilized to connect the terminal to a GPRS phone with BT card. The Basic Network Encapsulation Protocol(BNEP) functionality of the two devices BT protocol stack is used to seamlessly transport IP packets into BT data-link frames.

8. CONCLUSION

Several wireless access systems are today available in most big cities, which are frequently physically overlapping, yet belonging to separate administrative domains. Seamless mobility inside this areas is already possible but requires vertical handover capabilities at the user terminal. We proposed a new approach in designing vertical handover algorithms, which is not aimed at optimizing resource usage of the two integrated access network from their administrators' point of view, but should try to balance, from the end-user point

of view, the overall *cost* of vertical handovers with the actual benefits they bring to his actual networking needs. This approach doesn't lead to a single optimal handover decision function. Instead, it poses the problem to assess the impact of vertical handovers' times, as well as their frequency, with respect to the transport level protocols used by network applications. In this early work, we set-up an example scenario involving two of the most common radio access systems (GPRS and WiFi) and simulated network traffic of the main application types, being so able to contrast the handover strategy with its overall impact on the user's work session. We currently aim at extending this set of initial tests with more varied and even uncommon network scenarios and/or user mobility patterns, in order to derive more generally valid data characterizing this correlation. Specific optimal vertical handover strategies are misleading, if optimized from a single, network-level, point of view. It is of primary importance to model a realistic vertical handover-strategy vs user-satisfaction function, to be used by each mobile user to autonomously evaluate and apply, from time to time, the handover decision which is more convenient to his current network needs and actual mobility model.

9. REFERENCES

[1] http://www.bluetooth.com.

[2] A. Bakre and B. Badrinath. I-tcp : Indirect tcp for mobile hosts. In *ICDCS*, pages 136–143. IEEE, 1995.

[3] H. Balakrishnan, V. N. Padmanabhan, S. Seshan, and R. H. Katz. A comparison of mechanisms for improving tcp performance over wireless links. In *IEEE/ACM Trans. Networking*, volume 5, pages 756–769. December 1997.

[4] A. Calvagna, G. Morabito, and A. La Corte. Wifi bridge: Wireless mobility framework supporting session continuity. In *PERCOM*. IEEE, 2003.

[5] A.T. Campbell, A.G. Valko, and J. Gomez. Cellular ip. Internet Draft, 1998. draft-valko-cellularip-00.txt.

[6] Fan Du, Lionel M. Ni, and Abdol-Hossein Esfahanian. Hopover: A new handoff protocol for overlay networks. In *ICC2002*, pages 3234–3239, New York, May 2002. IEEE.

[7] George Edwards and Ravi Sankar. Microcellular handoff using fuzzy techniques. *Wireless Networks*, 4(5):401–409, August 1998.

[8] Gabor Fodor, Anders Eriksson, and Aimo Tuoriniemi. Providing quality of service in always best connected networks. *IEEE Communications Magazines*, 41(7), Jul 2003.

[9] T. Goff, J. Moronski, D. S. Phatak, and V. Gupta. Freeze-tcp : A true end-to-end tcp enhancement mechanism for mobile environments. In *INFOCOM*, pages 1537–1545. IEEE, 2000.

[10] Eva Gustaffson and Annika Jonsson. Aways best connected. *IEEE Wireless Communications*, Feb 2003.

[11] S.E. Kim and J. A. Copeland. Tcp for seamless vertical handoff in hybrid mobile data networks. In *GLOBECOM*, 2003.

[12] S. Mascolo, C. Casetti, M. Gerla, M.Y. Sanadidi, and R. Wang. Tcp westwood: bandwidth estimation for enhanced transport over wireless links. In *ACM MOBICOM*, pages 287–297, 2001.

[13] J. Mitola. The software radio architecture. *IEEE Communications Magazines*, 33(5):26–38, May 1995.

[14] M.Ylianttila, M. Pande, J. Mkel, and P. Mhnen. Optimization scheme for mobile users performing vertical handoffs between ieee 802.11 and gprs/edge networks. In *Global Telecommunications Conference*, volume 6, pages 3439 –3443, San Antonio, Texas, USA, 2001. IEEE.

[15] Senarath N.G. and Everitt D. Controlling handoff performance using signal strength prediction schemes and hysteresis algorithms for different shadowing environments. In *46th Vehicular Technology Conference*, pages 1510–1514, Atlanta, USA, April 1996. IEEE.

[16] T . Onel, E. Cayirci, and C. Ersoy. Application of fuzzy inference systems to the handoff decision algorithms in virtual cell layout based tactical communications systems. In *MILCOM 2002*. IEEE, October 2002.

[17] Hyo Soon Park, Sung Hoon Yoon, Tae Hyoun Kim, Jung Shin Park, Mi Sun Do, and Jai-Yong Lee. Vertical handoff procedure and algorithm between ieee802.11 wlan and cdma cellular network. In *CDMA International Conference*, pages 103–112, 2002.

[18] C. Perkins. Ip mobility support for ipv4. IETF RFC 3220, Jan 2002.

[19] Muhammed Salamah, Fatma Tansu, and Nabil Khalil. Buffering requirements for lossless vertical handoffs in wireless overlay networks. In *VTC*, Jeju, Korea, April 2003. IEEE.

[20] M. Stemm and R. H. Katz. Vertical handoffs in wireless overlay networks. *ACM Mobile Networking and Applications (MONET)*, 3(4):335–350, 1998.

[21] Z. Sun, L. Liang, C. Koong, H. Cruickshank, A Snchez, and C. Miguel. Internet qos measurement and traffic modelling. In *2nd International Conference on Conformance Testing and Interoperability (ATS-CONF 2003)*, January 2003.

[22] N. D. Tripathi, J. H. Reed, and H. F. VanLandingham. An adaptive direction biased fuzzy handoff algorithm with unified handoff selection criterion. In *Vehicular Technology Conference*, volume 1, pages 127–131, Ottawa, Canada, May 1998.

[23] UCB/LBNL/VINT. Network simulator - ns (version 2). software tool. www.isi.edu/nsnam/ns/.

[24] Erik Vanem, Stein Svaet, and Frederic Paint. Determining user satisfaction in a scenario with heterogeneous overlaywireless networks. In *Proceedings of IASTED International Conference on Communicationsand Computer Networks (CCN 2002)*, Cambridge, USA, November 2002.

[25] K.D. Wong and D. Cox. A handoff algorithm using pattern recognition. In *International Conference on Universal Personal Communications (ICUPC)*, pages 759–763, Firenze, Italy, October 1998. IEEE.

Author Index

NOTES